UNDERSTANDING
HOLOGRAPHY

UNDERSTANDING
HOLOGRAPHY

MICHAEL WENYON

ARCO PUBLISHING, INC.
New York

Second Arco Edition, First Printing

Published by Arco Publishing, Inc.
215 Park Avenue South, New York, N.Y. 10003

Library of Congress Cataloging in Publication Data
Wenyon, Michael.
  Understanding holography.

  Bibliography: p.
  Includes index.
  1. Holography. I. Title.
TA1540.W46   1985   621.36′75   84-14531
ISBN 0-668-06203-7 (paper edition)
ISBN 0-668-06414-5 (reference text)

Printed in the United States of America

10 9 8 7 6 5 4 3 2 1

Contents

List of Plates

'That window, that vast horizon, those black clouds, that raging sea, are all but a picture . . . You know that the rays of light, reflected from different bodies, form a picture, and paint the image reflected on all polished surfaces, for instance, on the retina of the eye, on water, and on glass. The elementary spirits have sought to fix these fleeting images; they have composed a subtle matter, very viscous and quick to harden and dry, by means of which a picture is formed in the twinkling of an eye. They coat a piece of glass with this matter, and hold it in front of the objects they wish to paint. The first effect of this canvas is similar to that of a mirror; one sees there all objects near and far, the image of which light can transmit. But what a glass cannot do, the canvas by means of its viscous matter, retains the images. The mirror represents the objects faithfully but retains them not; our canvas shows them with the same exactness and retains them all. This impression of the image is instantaneous, and the canvas is immediately carried away into some dark place. An hour later the impression is dry, and you have a picture the more valuable in that it cannot be imitated by art or destroyed by time . . . The correctness of the drawing, the truth of the expression, the stronger or weaker strokes, the gradation of shades, the rules of perspective, all these we leave to nature, who with a sure and never erring hand, draws upon our canvasses images which deceive the eye.'

Tiphaigne de la Roche, *Giphantie,* Paris, 1760
(English translation, *Giphantia,* London, 1761).

' I felt in touch with the modern world,
I felt in love with the modern world.'

Jonathan Richman and the Modern Lovers,
Roadrunner, © Modern Love Songs, 1975.
Used by kind permission of Beserkely Record
Company.

Introduction

Most dictionaries describe a holograph as a document written wholly by the person in whose name it appears. In the early 1960s US scientists, evidently unaware of the word's traditional meaning, used the word 'holograph' (or 'hologram') to describe a new device which produced a three dimensional image of an object; from *holo*, meaning whole, and *graph*, meaning picture. The principles of the technique of holography were first described by the British scientist, Dennis Gabor,* in 1948. Its importance was not fully appreciated at the time, and the few researchers who worked on it in the 1950s were hampered by the lack of a suitable source of light having the property known as coherence. In 1960 the first laser was produced; laser light has coherence, and its use enabled US scientists Emmet Leith and Juris Upatnieks to produce the first holograms depicting objects in three dimensions. Research continued throughout the 'sixties and 'seventies, and to date hundreds of scientific papers have been published about holography; many textbooks have been produced, though, on the whole, these appeal to the specialist rather than to the general reader.

Probably the simplest description of holography is 'three-dimensional photography with lasers'. It is not entirely satisfactory—there are many other forms of 3D photography—but it does convey many of the essential elements: holography is a technical process capable of forming a visual record of an object; the image produced is three-dimensional and appears as solid as the original object; and the use of lasers has had an important effect in facilitating its development. This book is concerned with what holography is, how it works, how it can and has been used, its limitations and particular properties. My

* In 1971, the Nobel Prize for Physics was awarded to Dennis Gabor in acknowledgement of the importance of his invention of holography.

aim has been to present these details assuming a minimum of previous acquaintance with those ideas of physics upon which the mechanism of holography is founded. In my own attempts to find out about the whole field of holography I was forced to search through a wide variety of sources, finding many of of them either over-specialised and technical or, at the opposite extreme, oversimplified and lacking in generality. While I cannot claim complete generality, I would like to think that this book covers, in a straightforward manner, most of the important topics in holography. Holography may be of use to artist and scientist alike, and I hope that this book may serve for either as a simple introduction to its use, while allowing the general reader to find out about the workings of this new technique.

A study of holography makes clear many of the assumptions we have in our attitude to ordinary photography. Holography may yet challenge the authority of photography to record faithful and intelligible impressions of the world around us. Occasionally, historians find it useful to divide the history of civilisation into epochs defined by the types of communication media prominent in a particular day and age; for instance, the use of hieroglyphs in ancient Egypt, or the invention of printing (1450). In recent times, as technology has become progressively more sophisticated, we have seen a host of new media introduced, such as the telephone and TV. Some would deny that these new media have had altogether beneficial consequences, but few would question the influence they have on life in the modern world. Though holography may still be in its infancy as a medium, we can reasonably expect that in the future it will increasingly replace, or at least extend, the use of existing, well established media. Despite this there is widespead ignorance of how holography works and the ways in which it can, and cannot, be used.

Accounts of holography in the popular press and in science fiction are often garbled, misleading and incorrect. At times they give an altogether mistaken impression of the possibilities of the technique. To see a hologram for the first time is a fascinating experience: I believe the physical explanation of the way it works to be equally so. Only then can one begin to appreciate both the potential and the limitations of holography, not just for the present but the future as well.

A note about terminology: from the nouns 'hologram' and 'holograph' (which have identical meaning) I have derived the verbs 'hologram' and 'holograph', which, though not in general usage, seem a logical extension in accordance with the common use of the word 'photograph'.

MW

Introduction to the Paperback Edition

Since this book was first published, many people have seen holograms for the first time, largely due to the increasing availability of mass-produced holograms. I have added new material for this paperback edition describing the technical and commercial developments which have allowed a greater realisation of holography's potential as a visual medium. Also, since many developments have come about simply because a greater number of individuals and companies now know how to make holograms, I have added a section about the teaching of holography and its use in education. It is interesting to note that many of the dramatic ideas outlined in the chapter on the future have yet to be realised! I have up-dated the sections on making holograms, and new books and articles are included in the bibliography.

MW
May 1984

1 About Light

What is Light?

What sort of answer do we expect to the question, 'What is Light?' A ship designer will investigate the properties of a particular hull by studying the behaviour of a scale model of the hull in a water tank; in a similar way scientists describe and investigate the world by making abstract models of it called theories. These theories are valid for as long as they continue to predict correctly the behaviour of the real world as revealed by experiments. In a sense, holography is a technique which was 'predicted' using theories about light by the British scientist Dennis Gabor in 1948. In order to understand how holography works we must describe these theories about the nature of light and its properties. This will be our answer to the question, 'What is Light?'

The theory that led Gabor to the discovery of holography is the **wave theory of light**, first suggested by Christopher Huygens (1629–95). We can interpret this theory as saying that light has certain properties which make its behaviour analogous with the waves on the surface of a pond. When we throw a stone into a pond a pattern of concentric rings of waves moves outwards from the point where the stone hits the water (fig. 1). These waves move outwards with a certain **velocity** and the

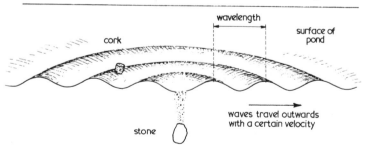

1 Waves produced by throwing a stone in a pond. Circular rings of waves expand from the point where the stone hits the water.

distance between each crest is called the **wavelength** of the waves. A cork floating on the surface of the pond will bob up and down at a particular **frequency** as the waves moving by it raise it up and lower it down. We say that the cork is **oscillating** due to the passage of the wave. In the interval between the passage of two successive wave crests we say that the cork completes one **cycle** of oscillation. The properties of velocity, wavelength and frequency are related to each other by the following equation:

$$velocity = frequency \ x \ wavelength.$$

With light, the colour of the light wave is described by either the frequency or the wavelength. Conventionally we specify the colour of a light wave by its wavelength. Red light has a wavelength of about $1/1,600,000$m while blue light has a wavelength of about $1/2,100,000$m. White light, such as light from the sun or from a light bulb, is a mixture of light of all different wavelengths; i.e., of all different colours. All light waves, regardless of colour, travel through space at the same velocity, about 300,000km/sec (186,000mi/sec).

As a light wave passes a point it produces electric and magnetic fields there. The strengths of these fields oscillate, just as the level of a cork does on the surface of a pond as a water wave passes. The strength of the fields at different points along the direction of the light wave is analogous to the height of a water wave along its length. This is why we represent the light as a wave travelling through space. It is not a wave describing the movement of a physical material as is the case with the water waves; it merely describes the way in which a quantity, the **field strength,** varies as the light wave passes a point.

Light is a type of electromagnetic radiation, as are radio waves and X-rays, and they all consist of oscillating electric and magnetic fields described by waves. All types of electromagnetic waves travel through space with the same velocity as light waves. These waves have a wide range of wavelengths, from rádio waves, which may be several metres long, to X-rays, which are one billionth of a metre long. They surround us at all times. 'Light' is the name we give to the narrow range of these wavelengths to which our eyes are sensitive.

We shall need to look closely at the behaviour of waves in general in order to explain holography. We may gain an intuitive grasp of many of the general properties of waves by realising that sound too is a wave phenomenon. The word 'wave' here describes variations in pressure to which our ears are sensitive. Sound waves are thus pressure waves. Waves on water, light and sound waves have many properties in common simply because they are all waves. Whereas the **colour** of a light wave is usually specified by its **wavelength,** the **pitch** of a sound wave or note is specified by its **frequency** (in cycles per second).

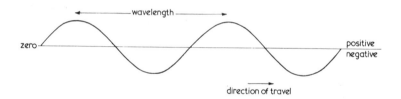

2 A sine wave. In the case of water waves, this would be a cross-section through the waves. With light waves this represents how the strength of the electromagnetic field varies along the length of the wave.

The simplest type of wave is known as a **sine wave** and is shown in fig. 2. Imagine this as a water, sound, or light wave travelling along in a certain direction. We have drawn the horizontal line to represent, in the case of water waves, the surface of the pond when it is still. The height of the wave is sometimes above this level, or 'positive', and sometimes below this level, or 'negative'.

In the case of a (sine) light wave the quantity analogous to the height of the cork is the field strength, which is sometimes

13

positive and sometimes negative. The dotted line in our diagram thus represents the strength of the field before the light wave passed; i.e., zero. The height of a wave is known as its **amplitude.**

All waves carry energy. It is one of their functions in the physical world to transmit energy from one point to another. The **intensity** of a wave is a measure of the amount of energy it is carrying. In the case of light waves this corresponds to the brightness of the light, and with sound waves it is the loudness (or volume). The intensity of a wave is simply the square of the maximum amplitude:

$$\text{intensity} = (\text{maximum amplitude})^2.$$

The sine wave shown in fig. 2 corresponds to light of one particular wavelength—i.e., one particular colour—or, in the case of sound, to a pure tone or note. Such perfectly pure colours and tones do not occur in nature though, as we shall see, they may be produced by artificial devices.

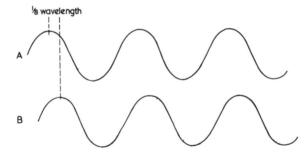

3 Sine waves with phase difference. We say that the phase of wave B is advanced by an eighth of a wavelength relative to that of A; i.e., the phase difference is an eighth of a wavelength.

Combining Two Waves

We now want to consider what happens when we combine two waves travelling together in the same direction, for instance the sine waves labelled A and B in fig. 3. We can see that both have the same wavelength: if they were light waves they would have the same colour. The maximum amplitude of both waves is the same; i.e., they have the same intensity. Notice that the crests of wave B are an eighth of a wavelength in front of the crests of wave A. We say that the **phase** of B is **advanced** by an eighth of a wavelength relative to A. If we 'held' wave A still and slid the wave B along we would obtain other phase differences between the two waves.

When we combine two waves we obtain a wave which is the result of adding the amplitude of wave A to the amplitude of wave B at each point along their lengths. Depending on the phase difference, one of the waves may be positive at any point while the other is negative. It is therefore possible to have a resultant amplitude of zero if the amplitudes of A and B happen to be equal and opposite.

For holography we shall be most interested in the two extreme cases; fig. 4a, in which the waves are **in phase**, that is they have their crests together; and fig. 4b, in which the waves are **in antiphase**, in which case the crests of A coincide with the troughs of B. For two waves which are in phase, the phase difference is zero, and for two in antiphase the phase difference is half a wavelength. Also shown in figs 4a and 4b are the waves produced by combining or adding waves A and B. When A and B are in phase they produce a bigger wave but when they are in antiphase they cancel each other out at all points. The phase difference of the waves A and B determines the size of the wave produced when they combine.

Thus if we add two identical waves which are in phase then we get a resulting wave of greater intensity—in the case of light waves, it would be brighter than each of the waves separately. If the waves are in antiphase they completely cancel each other out to give a zero intensity, which, if they were light waves, would result in darkness. The process by which two waves combine or add together to produce another is called **interference**. In common parlance this word implies some sort of destructive process but in the meaning we have defined it can be constructive as well: in the two cases just considered, combining waves in phase results in **constructive interference** and combining waves in antiphase results in **destructive interference.**

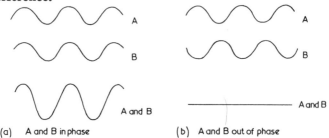

(a) A and B in phase   (b) A and B out of phase

4 Results of adding two sine waves: (a) when A and B are in phase—constructive interference produces a large wave—and (b) when the two waves are in antiphase—destructive interference results in there now being no wave at all.

Light Waves in Three Dimensions

The sine waves we have been drawing and talking about so far are waves which describe the way in which a quantity, such as the height of the surface of a pond, varies along a line in the direction of the wave. In the case of water waves this is just the cross section of the waves which are moving outwards from the centre where the stone hit the water.

The complete technical description of the waves produced by a stone thrown in a pond is that they consist of circular **wavefronts** expanding outwards from the point at which the stone hit the water. On any of these circular wavefronts the height or amplitude of the wave is the same.

Water waves occur on the flat two-dimensional surface of a pond. Light waves must be somewhat different since they travel through the three-dimensional volume of space. For example, a source of light which is a tiny point radiates light in all directions and the wavefronts proceeding from a point source of light are expanding spheres. Across the whole of the surface of one of these spheres the amplitude of the light waves, or the field strength, is the same. The amplitude on any

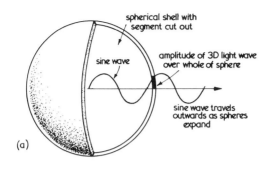

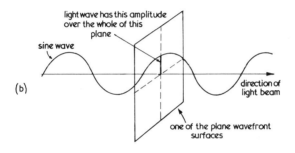

5 Simple light wavefronts: (a) spherical wavefronts proceeding from a point source of light, and (b) plane wavefronts in a parallel beam of light.

of the spheres is given by a sine wave which emerges from the point source and travels outwards with the spheres (fig. 5a). A point source of light is said to produce spherical light waves.

If light travels in a narrow parallel beam the appropriate wavefronts are planes which extend at right angles to the direction of the beam. The amplitude of the light waves on any of these planes is given by a sine wave travelling with the planes in the direction of the beam (fig. 5b). Such a beam is usually referred to as a plane wave of light.

In any given situation we may describe the way in which light waves propagate through space by specifying the shape

17

of their attendant wavefronts. The two examples we have given, of spherical and plane waves, have wavefronts with quite simple shapes. Also, since the amplitudes on the various wavefronts are described by a single sine wave, they must consist of light of one pure colour. They are, in fact, rather idealised examples of light waves. The light which is being reflected to you from this page, for instance, has waves and wavefronts which are complicated in a manner which we have yet to discuss. Firstly we shall consider what happens when we combine two simple light waves, since this has particular relevance to holography.

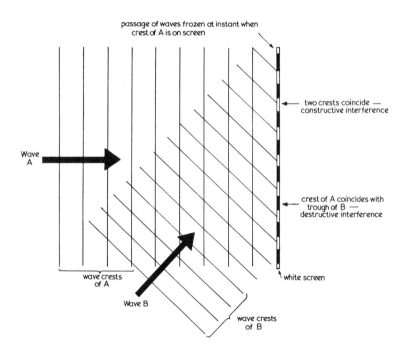

6 Interference of two plane waves on a screen (vertical view). The phase difference between the two waves varies as we move along the screen and we obtain, alternately, constructive and destructive interference.

Interference Patterns

What would happen if we simultaneously projected plane light waves of the same wavelength (colour) towards a screen from two slightly different directions? We might guess that the screen would be uniformly illuminated across its surface at a brighter level than if just one plane wavefront were projected at it. In fact this is wrong. At some points on the screen the waves are in phase and combine to produce bright illumination there, while at others the waves are in antiphase, cancel each other out and, as long as there is no other light hitting the screen, produce a dark region. The pattern of dark and bright lines produced on the screen is an **interference pattern**.

To see how this pattern is formed we must consider the detailed geometry of the two plane waves combining at an angle on the screen. In fig. 6 we are looking down on a cross-section of the two plane wavefronts hitting the screen. The straight parallel lines are drawn to represent the crests of the two sine waves which describe the amplitudes on the various plane sections of the two waves, and the passage of the waves has been frozen at the instant when a crest of wave A lies along the length of the screen. In between each crest of wave B there is of course a trough of negative amplitude.

We want to look at the type of interference produced at points along the length of the screen. As we said earlier this depends on the phase difference between the waves arriving at the point. Starting from one end of the screen and moving along it we see, by looking at the positions of crests and troughs of wave B relative to the crest of A, that the difference in phase between the two plane waves varies. At some points a crest of B coincides with the crest of A (i.e., the sine waves passing through these points are in phase), at other points wave A has its crest while wave B has a trough (i.e., the sine waves passing through these points are in antiphase). We therefore obtain at these points respectively constructive interference and destructive interference. By careful examination of fig. 6,

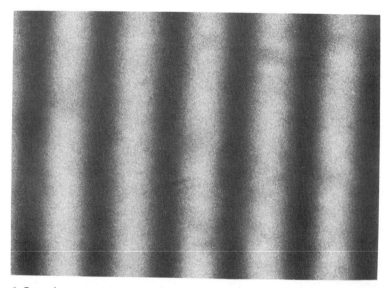

1 Interference pattern produced by two plane waves (magnification x 100). This pattern was produced by two parallel beams of laser light inclined at an angle of about 45°. It is unique to the particular configuration of the interfering wavefronts used.

we see that constructive interference and destructive interference occur alternately at regular intervals.

We have considered only one horizontal cross section of the plane waves; however, as we consider higher and lower cross sections we will get constructive and destructive interference at just the same points along the width of the screen. Whenever we get constructive interference we gain some intensity of light, and with destructive interference we get darkness. A pattern of alternate bright and dark parallel vertical lines is thus produced on the screen. Plate 1 is a photograph of such an interference pattern produced by the interference of two plane waves of light of one colour. (The fuzziness of these lines is due to the fact that our explanation here is a slight simplification: at points on the screen between the two extremes of destructive interference, giving zero intensity, and constructive interference, giving maximum intensity, there are regions of intermediate intensity where the brightness is gradually increasing and decreasing.)

The bright and dark lines which make up the interference pattern are known as **interference fringes.** The distance between the fringes in the pattern depends on the particular set-up used to make them. If the distance between wave crests is larger— i.e., if we use light of a longer wavelength—then the distance between the fringes will be larger too. So light of different colours produces patterns with different fringe spacings. If the angle between the two plane waves is narrower than in fig. 6, then this too increases the distance between fringes since the relative phase difference of the two waves changes more slowly as we move along the screen.

If we combine other pairs of simple waves on a screen we obtain different types of interference pattern. If light waves with a spherical wavefront (i.e., spherical waves) interfere with plane waves, then they produce on a screen an interference pattern which consists of concentric rings, alternately light and dark.

Although we have considered only waves with plane and spherical wavefronts, there are other types of simple waves represented by wavefronts. In general, any pair of these simple waves may combine at a screen to produce an interference pattern which is *unique* to that pair of waves and which depends only on the shape of the wavefronts which describe the light waves.

Black-and-white photographic plates and films are sensitive to light in such a way that, if a region of the plate is exposed to light, it becomes dark after the plate is developed, and will not allow light to pass through it. Any region of the plate not exposed to light becomes transparent after development. We can thus record the interference pattern of two waves by simply placing a photographic plate at the position of the screen, exposing it to the pattern of light and then developing it. The record we then have will be a photographic negative of the original pattern.

In order to see the interference pattern at first hand or to record it photographically there must be no stray light hitting the screen or photographic plate other than the waves we have projected at them. For this reason such experiments are carried out in a darkened room.

Natural Light

We showed in the previous section that the simple idealised plane and spherical light waves have the ability to produce interference patterns on a screen. In what sense are they 'simple' and 'idealised'?

We shall show that they may be produced from a laser using a variety of optical devices (such as lenses), but the laser (as we shall see) is a very special source of light. We might well ask why it is that light such as that produced by two electric lightbulbs cannot be used to produce interference patterns.

To answer this question we must look at a property of light known as **coherence.** Basically there are two types of coherence, **spatial coherence** and **temporal coherence,** and light waves from sources which do not possess a high degree of either of these will be incapable of producing interference patterns. Instead they will, on combining, produce a uniform illumination across the surface of a screen, devoid of any pattern of dark and bright fringes. Let us look at each of these types of coherence separately to see why this should be so.

For light to have **temporal coherence** it must consist of waves of a single well defined wavelength (colour); that is, it must be **monochromatic** light. We mentioned earlier that white light consists of a mixture of all colours. A parallel beam of white light would consist of a mixture of plane waves of all different colours travelling together through space in the same direction. Consider then what would happen if we projected two such parallel beams of white light towards a screen at different angles. Each of the component colours' plane waves would produce on the screen their own interference pattern of vertical fringes in that colour; for instance, the red light waves would

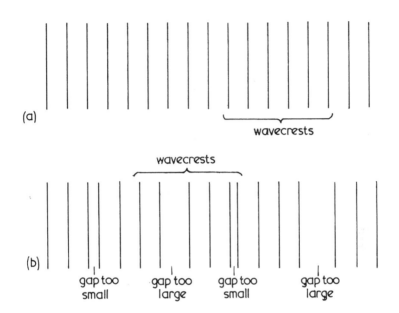

7 Temporal coherence: (a) good temporal coherence; (b) poor temporal coherence.

produce red fringes, and the blue light would produce blue ones. All these patterns become superimposed on the screen. However, since the distance between fringes is different for every colour, the patterns overlap each other on the screen and get mixed together. The result is a uniform white illumination of the surface of the screen in which no fringes are visible.

Another way of describing light with good temporal coherence is to say that all the wave crests must be 'in step'; i.e., they occur at regularly spaced intervals. In saying this we are likening the passage of, for example, a plane wave with a high degree of temporal coherence to the marching of a very long column of troops in which each row of soldiers is in step with the row before and after it (fig. 7(a) and (b)). If the crests of a plane light wave suddenly became 'out of step' with themselves (because of the appearance of an oversized gap between two successive crests), this would result in an equally sudden change in the phase difference between this wave and any other wave which is interfering with it. The interference pattern would then be shifted to either the left or the right on the screen.*

The average distance over which the wave crests remain 'in step' is known as the **coherence length** of the source which produced it. The longer the coherence length of the source, the more purely monochromatic it is and the easier it becomes to produce interference patterns with wavefronts derived from it. A source of light with a long coherence length has a high degree of temporal coherence.

To produce interference patterns we must use a source which produces light that is extremely pure in colour. Even if we filter white light from an electric bulb through a piece of red glass it would be difficult to produce interference patterns.

The best non-laser sources of monochromatic light usually have coherence lengths of less than one millimetre, whereas the coherence length of a laser may extend to one kilometre.

Spatial coherence describes the regularity of the phase of a light wave across its width (temporal coherence, as we have seen, is concerned with the regularity of the phase of a light wave along its length). A point source of light produces waves which have spatial coherence. The spherical and plane waves of fig. 5 both possess this quality of spatial coherence; the spherical waves have spatial coherence because they are, in fact, the type of wave produced by a point source of light; the plane waves have spatial coherence because it is possible, using a lens, to focus a perfectly parallel beam of plane waves to a

---

* In light which lacks temporal coherence the wave crests are not in step (there are occasional irregular spacings), and this shifting of the interference pattern occurs very rapidly in a random fashion. The net effect, as seen by our eye, is a uniform illumination of the screen. This is an alternative explanation to the one above of the inadequacy of natural white light to produce interference patterns. When one delves into the mathematical formulation of the theory of coherence the two are seen to be equivalent. We shall use either according to its applicability to a given situation.

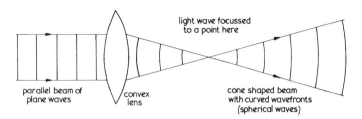

8 Plane wave focussed by a lens. A plane wave has spatial coherence since it can be focussed to one point by a lens. It proceeds from the point in a cone-shaped beam with curved wavefronts.

single point—once focussed to a point in this manner the waves proceed from the point as a cone-shaped beam of light, the wavefronts of which are curved, like the surface of a sphere as in fig. 8 (this is known as a diverging spherical wave or beam).

This is another reason why an ordinary electric lightbulb cannot be used to produce interference patterns. Because of its size it is obviously not a point source of light. One way to obtain spatial coherence in the light from a large source such as a lightbulb is to place it in a black box which has a small pinhole in one wall. Light which emerges from this pinhole has wavefronts which are hemispheres expanding from the hole. The fact that they are half, rather than whole, spheres is unimportant: the light still has spatial coherence. It is, however, not a very bright source since only a small fraction of the light produced by the bulb actually emerges through the pinhole. If we hoped to make interference patterns using this light, we would also have to filter it through a piece of coloured glass (to make it temporally coherent) and this would mean reducing the brightness even further. Any resulting interference pattern would be very dim indeed.

The laser is a device which produces light that has a high degree of both spatial and temporal coherence and, as such, is an ideal source of light to produce interference patterns which are bright and easily visible to the eye.

Seeing

Our eyes receive light from objects either because they are sources of light or because they reflect light. The reflection obtained can be of two types: either the surface is smooth like glass or a mirror, in which case we have **specular** reflection, or it is rough like wood or cotton, in which case we obtain **diffuse** reflection. At any time we may have both types occurring simultaneously, as in reflection from polished wood.

The light reflected from objects is in the form of complicated wavefronts which travel outwards in all directions. If we look at objects in a room lit only by daylight we are, in general, receiving waves of light which have neither spatial nor temporal coherence. If the objects are multicoloured or white then the waves must be 'multi-coloured'; i.e., they consist of a mixture of light waves of different wavelengths. Also, the shape of the wavefronts is quite complicated, though it is related to the shape of the objects.

But we don't 'see' waves, we see objects in the real world. In some way our eyes and brain use the light waves so that we see a three-dimensional coloured world around us. In an abstract sense the wavefronts of the light waves contain in their shape and structure information which our eyes and brain

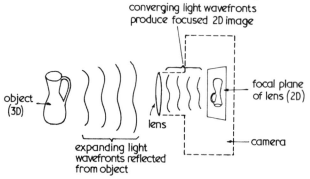

9 Action of camera lens producing a 2D image of a 3D object. The process by which a camera forms a flat image is one which we very often take for granted. The eye works in a similar way to the camera.

process to give us our view of the world. Exactly how this is done is a problem in experimental psychology; we shall be concerned only with describing the physical processes that occur in seeing.

The camera is a simple device which works in much the same way as the human eye. The camera has a lens which focusses the light waves expanding from an object to form a flat (two-dimensional) image at the **focal plane** of the lens (fig. 9). This can be recorded on light-sensitive film to produce a photographic negative which is then used in another process to produce a positive photographic print. By using a lens to focus the waves from an object we create a two-dimensional image of a three-dimensional object. This is, in fact, quite a remarkable process which we have come to take for granted since we see photographs so very often. Since the days of the earliest cave drawings people have been producing, with various degrees of success, two-dimensional representations of the three-dimensional world. Techniques have been invented or have evolved that have succeeded in doing this with an increasing sense of reality. 'Perspective' in paintings and drawings was perhaps one of the major discoveries. In a picture which uses perspective we can imagine lines in the picture which converge to a point in the distance; objects in the foregound are shown as being larger than objects behind them. The principle of perspective is seen in photographs and is responsible for their realistic portrayal of the world.

The human eye also is constructed with a lens. At the back of the eye is a light-sensitive surface called the **retina,** onto which an image is focussed in just the same way as in a camera. The retina may be regarded as a mosaic of tiny light receptors, called rods and cones, each of which sends messages to the brain *via* the optic nerve according to the brightness of the image at the particular receptor. They also send messages to the brain according to the colour of that part of the image. The messages arrive at the brain as a series of electrical currents or signals. All these signals are analysed by the brain in such a way that we see an object 'out there' in the world. This is where psychology comes in. We have two eyes and each sees a slightly different view of an object. By synthesizing these two views the brain can get information about the depth of objects in a scene.

For our purposes, rather than trying to explain exactly how we perceive the world as three-dimensional, let us say that our eyes receive from the world the light waves with complicated wavefronts that are reflected from objects and somehow, as a result, we perceive the three-dimensional world. In terms of our comments above about the two different views of our eyes, we may think of this as 'sampling' the light waves at two different positions. The brain then processes this information to give us our view of the world.

Certain things result from the way in which we perceive the world as three-dimensional. The most important of these is the phenomenon of parallax. If an object in the foreground of a scene is obscuring another behind it we simply move our heads to one side, thus 'looking around' the object in front to see the one behind. If you repeatedly move your head from side to side, objects in the foregound of a scene appear, in a sense, to 'move' more than those further away. Also, objects in the foreground appear to move in the opposite direction to those behind them. These effects are known as parallax.

A photograph, of course, does not show parallax. Once the photograph has been taken the relative positions of objects in the picture are frozen. A photograph does not reproduce the complicated wavefronts reflected from objects in a scene; it reproduces a flat image of the scene formed by a lens. This image is intelligible to our eye as a representation of the scene. We see the photograph by light reflected from its surface. There is a great difference between the light wavefronts reflected to our eyes from a two-dimensional photograph and those reflected to our eyes by three-dimensional objects.

Before moving to discuss lasers we should look at a well established type of three-dimensional photography, a system known as **stereoscopic photography.** This works on the principle, mentioned before, that much of our perception of depth in a scene is due to the fact that our two eyes produces dissimilar images on their retinas, which the brain converts to a kind of three-dimensional view. A stereoscopic viewer presents to each eye separately a photograph taken from the viewpoint that that eye would have had, had it been viewing the scene directly; these pairs of photographs are taken using a special camera with two lenses. On viewing the 'stereo pair' through a viewer we obtain the illusion of perceiving depth in the scene: it appears three-dimensional. An important limitation of the technique is that we obtain a view of the scene as if we were standing in only one particular position. It is not possible to look around foreground objects at those behind; i.e., the stereo viewer does not demonstrate parallax. Holographic images, as we shall see, show the full parallax effects observed in real life; for instance, we may look slowly and gradually to one side of an object by moving our head to one side, thereby seeing many different views of the object; an effect sometimes given the name **autostereoscopy.**

2 Production of Light: The Laser

With the invention of the laser in 1960, scientists received a new device which was to become a valuable tool. At first there was some uncertainty about the possible application for a device which could produce an intense beam of coherent light. But, needless to say, within a few years the laser had found countless uses in many branches of science and technology.

In fact, the theoretical principles of the production of light, upon which the invention of the laser was based, were understood well before 1960. They were formulated in the theory of Quantum Mechanics which was established during the first quarter of this century. The scientists concerned were inquiring into the means by which materials can, under certain circumstances, produce light or other electromagnetic radiation. They were also concerned with the effect of light on matter.

To explain the results of some experiments undertaken at the time, it was suggested that a beam of light consisted of a stream of many particles called **photons** which have no mass and which travel at the speed of light; this is known as the **quantal theory of light.** Although this may appear to contradict the earlier wave theory of light, scientists today are content to use either theory according to the circumstances under consideration. In a sense we can think of light waves as describing the effect of a stream of many photons, or, conversely, we may say that photons describe the effect of light waves.

Indeed we can associate with any photon a wavelength for light waves which 'travel with' the photon. An important part of the quantal theory of light is that each photon carries a definite quantity of energy according to the wavelength or frequency of the light waves associated with it. This energy is given by Planck's Law:

Energy of photon = Planck's constant x frequency of associated waves.

The energy in a light beam is thus the sum of many 'packets' of energy. The energy of each of these packets is very small. If a photon of red light gave up all its energy to a glass of water it would raise the temperature of the water by only about one billionth of one billionth of a Centigrade degree.

A photon can be absorbed or released by an atom. Atoms, the basic constituents of matter, may for our purposes be thought of as tiny billiard balls capable of absorbing or giving out energy. The atom may receive energy by colliding with another atom or with some other particle such as an electron or even a photon. When an atom absorbs energy we say that it is raised to an **excited state.** When this occurs the atom will return to its normal or **ground state** within a small fraction of a second. In so doing it produces a photon of light (or other electromagnetic radiation) which carries away the energy originally absorbed by the atom. According to Planck's Law, the frequency, and hence colour, of the light produced depends on the energy that the photon is carrying. So the colour of the light produced by an atom depends on the amount of energy it has absorbed.

All the different chemical elements in nature are each characterised by a different type of atom. It happens that an atom may not absorb just any quantity of energy. In fact any particular type of atom may absorb only one of a definite 'set' of different amounts of energy, and the different types of atom are characterised by the set of amounts of energy which they may absorb. For instance, an atom of one element may be able to absorb 1, 3, 5 or 9 units of energy, whereas an atom of another may absorb 2, 4, 8 or 12 units of energy. The atom which can absorb 1, 3, 5 or 9 units of energy is said to have four different excited states. Having absorbed one of these amounts of energy the atom then returns to its ground state and radiates a photon of light, the colour of which depends on the particular energy absorbed; i.e., on which excited state it was in. When an atom goes from an excited state to its ground state we say that it has made a **transition**. In the first example, since the atom may be excited to any of four higher energy states, there are four possible transitions back to the ground state, each of which results in a different colour being radiated.

Given a group of identical atoms we can supply them with energy in such a way that there will be at least some atoms in each of the excited states possible for that particular type of atom. Eventually they will all return to the ground state, and they will, as a group, radiate light that is a mixture of a small number of pure colours. For example, if the atom has a set of, say, four possible excited states then the group will radiate light which is a mixture of only four colours. We say that the **spectrum** of the atom consists of these four colours. Since every type of atom, or element, has a different set of excited energy states, each element has a unique spectrum containing a number of different colours. Indeed, astronomers study the spectra of stars in order to deduce the types of atom present in those stars.

Sources of Light

Light produced by heat

Atoms in any material are constantly in motion; e.g., in a solid they are vibrating about a fixed position and in a gas they 'fly' about freely, colliding with each other and with the walls of their container. As the temperature of the material is increased this motion becomes more violent. A particular atom will continually collide with other atoms and, each time this occurs, the atom will receive energy which will raise it to an excited state. It may then produce a photon of some sort of electromagnetic radiation. Objects around us are continually radiating electromagnetic waves but they are normally of very low intensity and of a wavelength to which our eyes are not sensitive. As the temperature of, say, an iron bar is increased, the atoms in it are excited to higher and higher energy states until they begin to emit photons of light.

At high temperature, however, another process, known as 'black body radiation', becomes active in any object. This produces light of all colours—i.e., white light—and effectively swamps the light produced by individual atoms. At a high temperature the iron bar will glow 'white hot'. In household electric lightbulbs, white light is produced by a similar process. The material of the bulb's filament is heated by passing an electric current through it so that it glows white hot. White light, as we mentioned in Chapter 1, does not possess temporal coherence; i.e., it is not of a single well defined wavelength.

Gas-discharge lamps

If our source of light consisted of atoms of only one type and we could excite them by some means other than heating, then the colours of light produced would be restricted to those in the spectrum of that single type of atom. By using even quite a crude filter we would be able to extract from this light just one of the colours in the spectrum. This would be very monochromatic and thus possess a high degree of temporal coherence.

The process outlined above occurs in a device known as the **gas-discharge lamp.** The neon light used in illuminated advertising signs is an example of this type of lamp. A simple version of this consists of a hollow, sealed glass tube containing neon gas (atoms of neon) at low pressure. Fused into the walls of the tube and protruding inside are two metal strips, called electrodes. If one connects the electrodes to a high-voltage direct-current electricity supply, one of them, the **cathode,** is made electrically negative in charge while the other, the **anode,** is made electrically positive. The cathode becomes negative because it has an excess of electrons, subatomic particles that have a negative electric charge. The anode is positive because it is deficient in electrons. The electrons on the cathode are

33

attracted to the anode (opposite charges attract) and literally fly off the cathode and through the gas to land on the anode. During their flight the electrons collide with some of the neon atoms and, in so doing, may give them energy. The neon atoms may accept this energy only if it is exactly that energy required to raise them to one of their permitted excited states. The electrons will have many different energies, some of which will be equal to energies required by neon atoms to raise them to their excited states; thus there will be some neon atoms in each of the permitted excited states at any time. As each atom returns to its ground state it will emit a photon of light, the colour of which depends on which state the particular atom was in. This process is happening continuously, with many atoms; as some atoms are emitting light, others are being raised to the various excited states by further collisions with electrons. A bright reddish glow appears in the region between the electrodes. The light being emitted actually consists of a mixture of light of a few discrete colours in the neon spectrum, but the red light is dominant. By passing the light through even a poor-quality red filter (i.e., one that is not particularly discriminating) it is possible to eliminate any other colours in the light and to single out the pure red light, which is very monochromatic. The filter allows us to extract just one colour (red) from the spectrum of colours of the neon atoms.

We may obtain light of other colours in a similar way by filling the tube with a gas other than neon. For instance, a discharge lamp using the gas argon produces a mixture of the colours in its spectrum which appears blue-green.

By filtering just one colour from the spectrum of colours produced by identical atoms, the gas-discharge lamp produces light with a high degree of temporal coherence. Interference patterns may be produced with light waves derived from the filtered light of gas-discharge lamps. In order to ensure that the light is spatially coherent we must use only that light which passes through a single pinhole. The patterns are consequently quite dim.

In fact the colour of the light produced by a discharge lamp with a filter is not perfectly pure. It produces a small range of colours which are almost, but not quite, of the same colour or wavelength. In practice, discharge lamps rarely have a coherence length of more than a few millimetres. As we shall see this places strict limitations on their use as sources of light in holography.

The lack of purity of colour in the light produced by a discharge lamp is due primarily to an effect known as the **Doppler shift.** Neon atoms, and indeed atoms in any gas, are in constant random motion. They move about in all directions, colliding with each other and with the walls of the tube; when an individual atom happens to emit light it may be moving either towards us or away from us at some considerable velocity. You may have noticed how the pitch of an ambulance siren depends on whether the ambulance is travelling towards you or away from you when it emits its note; this is a demonstration of the Doppler shift with sound waves. In an entirely similar way the colour of light waves emitted from an individual atom when they are received by us depends on whether the atom is moving towards or away from us when it emits the light. The combined effect of all the neon atoms which are moving and emitting light photons is to produce a small range of slightly different colours of light, all of which we would judge as 'red'; i.e., slightly different shades of red. It is because of the existence in the light from neon atoms of these slightly different, but nearly identical, colours that such lamps rarely have a coherence length of more than a few millimetres. In any conventional source of light (including those using solid materials rather than gases) the atoms are moving; as a result we always get this impurity in the colour of light.

In seeking a source with a high degree of temporal coherence it is necessary to conceive a system which can produce light of a colour which is independent of the motions of individual atoms. If at the same time we can contrive to make each atom produce light waves which are in phase with those of other atoms, instead of being independent, then we will have created a device which has a high degree of not only temporal coherence but of spatial coherence as well.

The Laser

In a neon lamp, individual neon atoms emit a photon shortly after receiving energy from a colliding electron. Each atom emits a photon of light spontaneously and independently of any of the other atoms around it. This process is thus called **spontaneous emission**, to distinguish it from the process by which photons are produced in a laser, which is called **stimulated emission**. We may think of an atom in an excited state as 'storing' a photon of light of a particular colour. Stimulated emission occurs when such an atom is hit by a photon with the same colour as the stored one. The stored photon is instantly released and it travels away from the atom accompanied by the colliding photon, which survives the impact. So, as a result of this collision, from one photon we now have two photons; but, most important, the waves of both these photons are in phase.

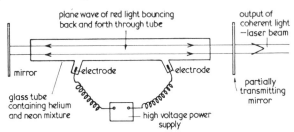

10 The helium-neon laser. This laser is often used in holography because it is relatively cheap and has good coherence properties.

As we shall see, the process of stimulated emission is crucial to the working of the laser. It was first understood by Albert Einstein at the beginning of this century. In fact, it does occur in a neon lamp but the effect is insignificant since it happens

36

only rarely. This is because relatively few of the neon atoms are in the excited state at any one time, and it is unlikely that any of these will be hit by a photon. The effect would become more significant if we could arrange for there always to be a majority of neon atoms in the excited state, but this would never occur under normal conditions; to achieve such a situation we must continually 'pump' large amounts of energy into the group of neon atoms.

We shall describe a particular type of laser called the helium neon gas laser. This consists of a long thin glass tube with electrodes, containing a mixture of the gases helium and neon (fig. 10). At either end of the tube are mirrors accurately aligned so that they cause light to be reflected back and forth through the tube. In order to make this device work as a laser we must attain the situation described above in which there is always a majority of the neon atoms in the excited state. As in the neon lamp, the neon atoms are raised to the excited state by passing an electric current between the electrodes and so creating a flow of electrons through the gas. Helium is mixed with the neon because it increases the efficiency of this process and enables a majority of the neon atoms to become excited.

Once this situation has been attained, some photons will inevitably be produced by spontaneous emission from the neon atoms, and will radiate in all different directions. As these photons travel through the gas they will collide with excited atoms and stimulate the emission of other photons in a chain reaction. After the first collision we have two photons; each of these will stimulate the emission of another photon, to give a total of four; these four will stimulate four more, to give a total of eight; and so on until the photons leave the glass tube. All of the photons will be in phase with the original photon.

If this were the end of the story nothing particularly remark-able would have happened; in fact, the net effect would be rather similar to the normal neon lamp, the light produced being not perfectly pure in colour (i.e., not temporally coherent) since the colour of the light emitted by the atoms is still affected by the Doppler shift. Neither is the light spatially coherent. However, some of the photons that are spontaneously produced—and it need be only a few of the many thus emitted—will happen to be travelling along the length of the glass tube. As they do so they will stimulate the emission of other photons, some of which will also travel along the tube. When these photons leave the end of the glass tube they hit one of the mirrors and are reflected back along the length of the tube for a second performance, stimulating the emission of more photons, which may join them. It would seem that the photons could bounce back and forth between the mirrors indefinitely, continually producing more photons. Remember, however, that at this point the light waves being generated in the laser are of several different colours (or wavelengths). To understand what does in fact happen we must go back to the wave theory and consider what happens to light waves reflected back and forth between two flat mirrors.

The two flat mirrors and the space between them, whether it contains air or the helium and neon gases or is just a vacuum, are collectively called an **optical resonator**. The phenomenon of resonance is more commonly encountered with sound waves than with light. An object such as a wine glass has a few particular frequencies at which it vibrates most easily; in fact, if we direct at the glass a note whose pitch is one of these resonant frequencies, we can cause it to vibrate so much that it will shatter. Some opera singers are reputed to be able to shatter a wine glass by singing a note of the correct pitch (or frequency). In an optical resonator we have light waves reflecting back and forth between the mirrors rather than sound waves making an object vibrate. We shall consider what happens to light of different colours, or wavelengths, rather than sound of different frequencies. Light waves reflecting back and forth between the mirrors will be in the form of plane waves. In a laser, the light waves of the different shades of red produced by the neon atoms (due to the Doppler shift) will be plane waves of various wavelengths. The optical resonator will 'resonate', however, for only one of these various wavelengths; by 'resonate' we mean that only light of this precisely specified wavelength (or colour) will build up in the laser and continue to be reflected back and forth between the mirrors.

In the helium neon laser, although plane waves of various shades of red may begin to bounce back and forth between the mirrors, the properties of the optical resonator ensure that only a plane wave of a very precisely defined colour will build up inside the laser; such a wave will, in fact, thrive, for it will stimulate the emission of other photons (i.e., more light waves) which will be in phase with the wave and may even be travelling in the same direction, in which case they immediately join the wave, and contribute more energy to it as they do so. So the plane wave which builds up in the laser is not only temporally coherent, being of a pure colour, but also spatially coherent, since it is a perfect plane wave.

An output of coherent light is obtained from the laser by 'tapping off' some of the light in the plane wave it produces. We may do this by making one of the mirrors so that it transmits (i.e., 'lets through') some of the light as well as reflecting some. From this end of the laser we then obtain a beam of coherent light. The beam is very thin when it emerges, typically a few millimetres in diameter, and usually appears as a glowing solid 'rod' of red light as it illuminates particles of atmospheric dust and smoke in its path. The beam travels in a straight line and hardly spreads out at all as it does so—it is possible to produce a laser beam which when pointed at the Moon (400,000km away), produces a disc of light on the Moon's surface only 3km in diameter.

The light produced by a laser has a high degree of both temporal and spatial coherence. The coherence length of lasers varies from about half a metre up to one kilometre, according to quality and expense.

The laser was preceded in its development by a device called the **maser**, which was invented in the 1950s. Similar in principle to the laser, this device produced coherent microwave radiation (electromagnetic waves longer in wavelength than light and classed as a type of radio wave). It was in 1958 that the US physicists Charles H. Townes and Arthur L. Schawlow published a paper on infrared and optical masers. In this they described the design of a laser and the principles of laser action. The first laser was made to work by Theodore H. Maiman in July 1960; he was working at the Hughes Aircraft Company's research laboratory in Malibu, California.

The first laser was a ruby laser. In this the 'lasing medium', instead of being a mixture of helium and neon gases in a glass tube, was a solid rod of synthetic ruby—this type of laser is known as a solid-state laser, as opposed to a gas laser; in general its output is less coherent but more powerful than that of a gas laser. Solid-state lasers also tend to produce light in short bursts or pulses whereas gas lasers give a continuous output of light.

Since 1960, lasers have been produced with many different solid and gas lasing mediums, each producing light of a different colour, power and coherence. Semiconductor lasers made from the materials used in transistors have also been produced.

The helium-neon laser we have described is commonly used in holography. A power output from the laser of a few thousandths of a watt is sufficient. This is very small and should the beam strike your hand it would not feel hot and would not burn or damage your skin. However, looking directly into the beam is very dangerous, since the lens of the eye will focus the beam to a tiny spot on the retina which may thus become damaged. It is necessary when working with lasers to take great care to avoid this.

For holography it is necessary to spread out the thin beam of the laser so as to illuminate the whole of the object being holographed. This may be accomplished using a lens known as a beam spreader, which converts the narrow straight beam to a cone-shaped beam that spreads out from the lens to provide a disc of illumination (as in fig. 8). An object illuminated in this way by a laser appears to have a peculiar grainy pattern across its surface. This is, in fact, a consequence of the high coherence of the laser light, and is called **laser speckle**.

3 Applications of Lasers

Dr Maiman, who made the world's first working laser, called it a multi-million dollar 'solution in search of a problem'. That was in 1960. Within eighteen months over 400 different companies were active in various types of laser research. Today the laser has found such a wide variety of applications that whole books are devoted to specialised aspects of these applications. We shall try to demonstrate briefly some of the wide-ranging applications the laser has found in today's technology.

High Power Lasers

Lasers can be made which have extremely high power outputs. So-called 'gas dynamic' lasers have been produced which have continuous outputs of hundreds of kilowatts concentrated in their narrow beams. As well as this, the coherence of laser light means that we can use a lens to focus all this power to a single tiny spot, thereby attaining temperatures in excess of 30,000°C. At this temperature any material will be vaporised so that the laser can be used as a cutting tool which can penetrate every material—including diamond. (There is a story that in the early days of laser development a research group was looking for a rough and ready measure of the power output of their laser under various conditions. They invented a unit called the 'gillette', which was the number of razor blades the focussed beam of their laser could penetrate.)

The laser's focussed beam can be made to follow a particular line across the surface of, say, a thin sheet of metal and so be used as an efficient cutter. It has found application in this way for cutting out metal parts for ships and pieces of fabric for suits. It has the advantage of being very accurate, as the beam may be focussed to produce a spot of light 0.0025mm wide. Lasers are finding considerable application in surgery. They have been used for some time to spotweld detached retinas to

2 Laser Knife. This knife, developed for general surgery, seals off blood vessels as it cuts, reducing blood loss by as much as 90%.

the back of the eye. A laser knife has also been developed for use in general surgery (plate 2). Held in the surgeon's hand this produces a very fine, clean cut and has the added advantage of sealing off any blood vessels that it severs, thus reducing the amount of blood lost by up to 90%; this makes the knife particularly useful for operations on haemophiliacs and for those parts of the body where blood vessels are very numerous, such as the tongue and the kidneys.

At the high temperatures produced by a laser beam it is possible to make certain nuclear reactions occur which would be impossible to produce otherwise. At very high energies it should be possible to cause the nuclei of atoms such as hydrogen to fuse or coalesce. When this occurs, much more energy is released than is used to make the nuclei fuse. This technique of **laser fusion** has the potential to provide us with an unlimited source of power. The 'fuel' for this reactor would be atoms of elements similar to hydrogen; these may be obtained by an extraction process from seawater, and are thus in plentiful supply. Generation of electricity by this process would require a new form of nuclear reactor; the problems are similar to those that already exist using the process of nuclear fission, but the rewards of having an unlimited energy supply—which would help to alleviate the 'energy problem'—are very high, and millions of dollars have been spent on developing the process. In 1976 the world's most powerful laser fusion system, the 'Argus' at the Lawrence Livermore Laboratory in California came 'on-line'. Each arm of the laser is capable of producing a short pulse of light the power of which is equivalent to the total output of the USA's combined electrical generating capacity (i.e., one terrawatt = a million million watts). In the UK, too, attempts are being made to harness the energy produced by laser fusion. (Laser fusion, as we shall shortly see, has more sinister applications in defence and thus much work on it is top secret.)

There are a variety of other applications for the high-power lasers. In Australia a lighthouse has been built equipped with a laser that provides a high-intensity beam which can be seen many miles away, even in poor weather. It has been suggested that lasers could be used as a method of clearing oil slicks at sea; a joint international authority could patrol the seas in ships equipped with nuclear powered carbon dioxide lasers which would be capable of vaporising oil on the sea's surface.

Lasers in Communications

In scientific terms, communication is the transfer of information from one place to another by any means, such as radio waves, telephone wires or morse code with light. The information may be transmitted by a human voice, a television picture or the printed word sent by teletype. Communications engineers are concerned with maximising the amount of information which may be efficiently transmitted by a given means.

To date the primary means by which information, known to the engineer as a 'signal', has been transferred from A to B has been either as some form of radio wave or by electric cable. Several signals—e.g., a number of different telephone conversations—may be carried simultaneously by a single radio wave. To achieve this the conversations or 'signals' are combined and converted into a special code and the radio wave carries the information in this encoded form. At the receiving end the radio wave is decoded; i.e., separated into the different telephone conversations which are then independently routed by cable to their appropriate destinations.

However, there is a definite limit to the number of telephone conversations (or the amount of information) which a single radio wave may carry. Using radio waves of shorter wavelengths it is possible to increase the information transmitted by a single wave. There is thus an obvious advantage in using radio waves of the shortest possible wavelength. Radio waves are a type of electromagnetic wave, like light waves, and differ from light only by their means of production and by the fact that they have much longer wavelengths: if we could transmit signals on light waves in the same manner as has been done with radio waves, then we could further increase the amount of information which may be transmitted on one electromagnetic wave; theoretically, in one light wave we could transmit 100,000 times more information than with the shortest of radio waves.

All this has been known for many years, but the implementation of communication by light wave has awaited the solution of a number of practical problems. Firstly, such a system requires a coherent source of light; with the invention of the laser this problem was solved. Secondly, light, unlike radio waves, is unable to penetrate fog, mist or cloud and cannot

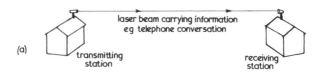

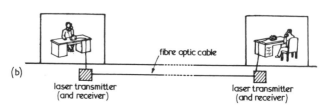

11 Laser-beam communications system: (a) communication by direct transmission of laser beam, (b) communication through fiber optic cable.

bend around corners. Early light communication systems, such as one built in the Kungei Ala-Tau mountains in Russia, were limited to fair-weather use between two stations visible from each other; i.e., on a line-of-sight.

This latter problem has been solved by the invention of a device known as a **fiber-optic,** which acts as a light-bearing pipe. A fiber-optic is a hair-thin strand of glass which transmits light by bouncing it off its walls and along its length. It is flexible enough to be bent around corners. Twenty-four such fibres may be combined to form a 'cable' which, including a protective covering, is only 1cm thick. If the laser light signal is to be transmitted for a long distance through the fibre, then the glass used to make the fibre must be of an extremely high purity and absolutely clear, otherwise the light signal could not penetrate the entire length of the cable. The glass used is so transparent that if the water in the deepest ocean were as clear it would be possible to see to the bottom of it.

Several experimental telephone systems have been set up with laser 'transmitters' and appropriate receivers, in which the fiber-optic cable is routed through the underground cable ducts of conventional telephone systems. Through the centimetre-thick light-cable described above it is possible to convey simultaneously over 8,000 telephone conversations. This represents a considerable saving in space over wire cable systems—an important consideration since the underground ducts in large cities are fast becoming overcrowded. There are other advantages: glass as a raw material is more abundant and cheaper that the copper used in conventional cables; the system is not prone to 'interference' from nearby electricity cables; and such a system can carry enough information to make video telephones a possibility at some time in the future.

The fact that a laser beam can carry a human voice has produced a novel application in the world of espionage. The sound of two people having a conversation in a room causes any windows in the room to vibrate slightly. If a laser beam is directed at the window and its reflection captured by a sensitive light detector it is possible to convert the movements of the window back into sound; i.e., into the original conversation, which can then be heard over a loudspeaker by enemy agents.

Metrology

Metrology is the science of measurement; it is concerned also with the accurate alignment of the various parts of a structure which is being built. The increasing sophistication of modern engineering demands that the components of a machine such as a jet aircraft be precisely positioned on assembly. The dimensions of aircraft now being built must be correct to within 0.5mm over a distance of 200m. The straight, narrow beam of the laser, acting as a convenient 'plumb line', makes such accurate alignment possible. It has been used also to ensure the straightness of ditches dug for pipelines.

Radar also can be considered a branch of metrology. Conventional radar is used to determine the distance of remote objects from a radar station. A single short pulse of radio waves is sent out from the station, hits the object, and is reflected back to the station where a receiver picks up its return. By measuring the time taken for the radio wave to make the journey, and knowing its velocity (which is the same as that of light), it is a simple matter to calculate the distance of the object. This may be done automatically by computer and displayed on a radar screen. Exactly the same process may be accomplished using light waves which, because they are of a shorter wavelength, produce a more accurate result. A laser beam aircraft tracker, working on the radar principle, is reputed to be able to determine the position of an aircraft 100km away to an accuracy of 10cm. Using the narrow beam of the laser in such a light radar we can be very selective and measure the distance of just one small object out of a group of several neighbouring objects. This technique is known as **laser-ranging**, and has obvious application for military purposes in determining the precise positions of targets.

Among its more peaceful purposes, laser-ranging has been used to establish the exact distance of the Moon from the Earth. As the Moon travels through its orbit this distance varies in a way which should be correctly predicted by various proposed theories of gravity, such as Einstein's General Relativity Theory. Measurement in the variation of the distance by laser radar thus provides us with a test of the different theories. It is made easier by the fact that one of the Apollo missions to the Moon left a mirror behind pointing at the Earth for just such a purpose. The measurements were made and Einstein's Theory survived the test.

By scanning across the surface of the Moon or a planet such as Venus with a laser radar in an orbiting satellite it is possible to build up a crude contour map of the surface below. Such a contour map would be extremely useful in planning the landing sites of future manned exploration of the planets and their moons in our Solar System.

Lasers have been used also to help in the prediction of earthquakes. The surface of the Earth (including the beds of the oceans) is divided into huge plates. These are very slowly drifting in different directions; they collide with each other, and earthquake and volcanic regions are pronounced along the borders between them. Earthquakes occur when the pressure built up along the border between two colliding plates is suddenly released by a jarring slip of one over the other. The continents, on different plates, are slowly drifting towards or away from each other. The rate of drift is very small, about 1 to 6cm a year.

In May 1976, NASA launched the laser geodynamic satellite, Lageos, to measure this drift and monitor displacements along the San Andreas Fault in California using laser radar. The San Andreas Fault is on the border of two plates and movements along it are the cause of earthquakes in California. It is hoped that predictions may be made using the Lageos satellite to tell of an impending major earthquake.

Lasers in War

A year before the first laser was made to work in 1960, the American Defence Department was negotiating with the Hughes Aircraft Company on the military applications of the device being developed in their laboratories. Since then, the main financial support for laser research has always been for work orientated towards its military applications, although these receive comparatively little mention in the public media. The US Air Force magazine said in January 1972 that laser weaponry is 'one of the most closely guarded national secrets'. In fact, more lasers are now sold for military purposes than for all other purposes combined. When the laser was first developed it seemed that it would become the 'death ray' of the science-fiction writers, concentrating kilowatts of radiative energy in its narrow beam and vaporising anything in its path. It was not until 1968, however, that lasers which even approximate to this description were produced.

The first military application of the laser was in the guidance of bombs, missiles and artillery shells. A front-line observer using a laser places a spot of laser light on the target (fig. 12). The missile is equipped to 'home in' on this spot using detectors sensitive to the laser light. It can do this with devastating accuracy: a bomb will land within 4m of its target; an artillery shell may be made to hit an object 2m wide 30km away. Laser guided bombs were first used in Vietnam in April 1972 and they made a big difference to the success of B52 bombing missions.

Eventually lasers were developed powerful enough to be classed as 'death-rays'. In June 1971 a laboratory was built near Alamogordo, New Mexico (where the first atomic bomb had been tested in 1945), to test gas lasers having an output of hundreds of kilowatts. Using these it is possible to set fire to wooden targets over 3km away.

As a weapon the laser has many advantages over conventional weapons. Its beam travels at the speed of light, and it is hoped that it could be used for knocking out aircraft and missiles. The possibility exists for launching satellites armed with such lasers which could hit almost any target on the surface of Earth. Laser beams have been suggested as a way of 'shooting down' hostile satellites.

Laser fusion, as we have seen, could be a means of supplying us with electrical energy in fusion reactors. It may also be

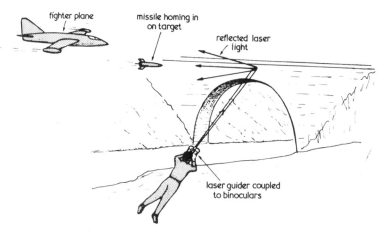

12 Laser guidance of missiles. A ground-based observer places a spot of laser light on the target. The aircraft fires a missile equipped to home in on the spot.

important as a means of triggering a fusion bomb. Like many other branches of laser research this area is top secret.

In discussions on arms limitations great emphasis is laid on the danger of the proliferation of nuclear weapons. It seems, however, that the potential destructive capacity of laser weapons systems being developed may be such as to render the nuclear threat comparatively insignificant.

In the editor's introduction to the 1976 edition of *Jane's Weapons Systems* (published by Jane's Yearbook), Ronald Pretty states:

> Beneath the guarded references to high energy laser research and development in American Department of Defense publications, and behind the virtual Soviet silence on the subject, it is probable that these two powers are locked in a costly super-scientific struggle to be first with a practical laser weapon capable of destroying a military target . . . solely by means of the energy the laser is able to generate, and transmit to the target, in fact the 'death ray' so beloved of generations of fiction writers.

Professor A. L. Schawlow, who in 1958 published the original paper on optical masers, has given lectures in the USA on the applications of lasers, omitting to mention the military applications. When questioned at a meeting about this omission he replied, 'I don't know what the military applications are; I don't want to know.' It is a dilemma of our times that almost any technological innovation has military as well as peaceful applications. The question of how much responsibility individual scientists should take or are even capable of taking, for such harmful applications, has long been the subject of heated debate in the scientific community (Barkan—see Bibliography).

Laser Light Shows

Many artists producing special-effect light shows have been attracted to the laser as a source of light because of its special properties and its brightness. Lasers have been developed that produce several colours, such as the krypton-argon ion laser which emits blue, green and red light simultaneously. There are several techniques used in laser light shows. The single beam may be split into a **fan** of many identical beams which can be reflected from mirrors to form a network of criss-crossed laser beams over the heads of the audience. Dust particles in the path of the beams are lit up by them so that the beams appear as seemingly solid rods of radiant light. This network, which is made to move while the audience watches, may consist of laser beams of many different colours.

Another technique is scanning. The thin beam of the laser is made to reflect off a tiny mirror onto a screen (or even on to clouds) where it produces a spot of light. The mirror may be made to tilt in any direction by a sensitive electrical servo, causing the reflected laser beam to turn with it. If this happens a line is traced on the screen. Using a magnetic tape made with the help of a computer we may send to the servo the appropriate electrical signals to move the mirror and cause the beam to trace out on the screen a line drawing of an object or some abstract pattern. This tracing, or 'scanning', is done many times each second and the spot of light moves so quickly that we cannot see it move; all we see is the traced out image produced on the screen. Animated pictures may be produced by projecting a sequence of these images, as in a cartoon. It has been suggested that huge animated advertisements could be simply projected onto large billboards acting as screens. Attempts to project images into clouds—'laser sky writing'—have not been entirely satisfactory.

Another technique used may be called **random phase modulation**. A laser beam is passed through a piece of glass or crystal which has an irregular, undulating surface. The light is spread out and bent in many different directions to produce a beautiful pattern when projected onto a screen. If the piece of glass is moved, the complex abstract pattern changes. At times it may appear to be like a huge spider's

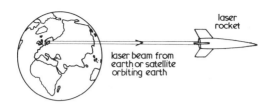

13 Laser rocket propulsion. This scheme could do much to lessen the expense of future space missions.

web and at others like the complicated star pattern of a galaxy seen through a telescope.

Laser light-shows have not only been used to dramatise the performances of pop groups; several have been produced as shows in their own right. The venues for these are either converted cinemas with special screens added or planetaria, such as the one in London, which allow the use of the dome ceiling screen. Theatres designed specifically for laser shows have been suggested. One firm markets a 'home' laser with accessories (a phase modulating device), enabling one to enjoy a laser light show in the comfort of one's own living room.

Other Applications

A phenomenon known as radiation pressure has been demonstrated with lasers. As we mentioned, light may be thought of as a stream of particles called photons. When these photons hit an object they exert a force on it. So a beam of light shone on an object exerts a pressure, known as **radiation pressure**. Scientists have succeeded in using this force to levitate a tiny glass bead 0.001mm in diameter. Although slightly larger objects could theoretically be levitated using higher-power lasers, these would tend to heat and possibly damage the object. Levitation of everyday objects is a practical impossibility by this means.

Laser beams carry energy. Theoretically, it should be possible to use laser beams like electric pylons to supply distant places with energy. This is not practicable for the national grid, but could be very useful as a method of beaming energy to spacecraft and space stations of the future.

The use of high-power lasers to beam energy to power rocket motors on aircraft and spacecraft is also being investigated, both in the USA by NASA and in the USSR (fig. 13). The technique had also been suggested as a means of providing propulsion for spacecraft in 'deep space'; far away from the Earth at the edges of our Solar System or beyond; as has the 'laser sail', where lasers would be directed onto a very light but very large sail attached to the payload: the small acceleration given to the sail by the impacts of photons would, over a few years, result in an exceptionally high velocity— theoretically, close to that of light itself. Yet another suggestion is that unmanned aircraft powered by laser rocket motors could be used to lift a transmitting aerial for a television station to 30,000m, resulting in a vast increase in the range of the station.

There are other applications of lasers not mentioned here, such as their use in pure science to explore the fundamental nature of solids, liquids and gases. In future there are bound to be many more. One thing is quite certain: there has been and will be no shortage of applications for this 'solution without a problem'.

4 Principles of Holography

In Chapter 1 we said that light consists of waves. We see an object as a result of the light waves reflected from it arriving at our eyes. The light waves reflected from an object are specified by describing the shape of a surface known as the **wavefront** which is uniquely related to the shape of the object. Two sets of coherent light waves may interfere to produce a pattern of bright and dark lines, or fringes, which is unique to that particular combination and depends only on the shapes of the two wavefronts which describe the waves. Such a pattern, known as an interference pattern, may be recorded on a photographic place placed in the region where the two waves are combining or interfering.

In the following discussion we shall often speak loosely of **wavefronts of light** (or wavefronts) when in more precise terms we mean 'a set of light waves described by a wavefront of such-and-such a shape'. This is merely a convenient abbreviation and is consistent with the terminology used by holographers themselves. For example, we shall talk of a 'plane wavefront of light' or just a 'plane wave front' instead of the more long-winded phrase 'light waves whose wavefronts are planes'. In fact, in this simple example, we could use the phrase 'plane waves'; however, the light waves reflected from an object cannot be described in such a succinct fashion, since the wavefront is in general of a complicated shape. In the case of light waves reflected from an object, using this new terminology, we speak simply of 'an object's reflected wavefront', thus obviating the need to describe the precise shape of the wavefront.

Holography is a means by which we may record the wavefront (i.e., the light waves) reflected from an object and, later, reconstruct this wavefront in such a way that the effect on an observer is to produce the sensation of actually seeing the object. It follows that the 'image' produced is three-dimensional, in just the same way that real objects appear three-dimensional. We shall have more to say about this later.

There is such a large variety of different types of hologram that it would be easy to become confused. In fact, to ascribe unambiguously a name to any particular type of hologram would require the use of four or five adjectives. The different basic types of hologram are presented here in broad and very general categories which are by no means exclusive. The logic behind the various classifications may not, at first sight, appear to be particularly clear; often a distinction is made due to some subtle point of theory.

Before discussing holography itself a word must be said about a wave phenomenon known as **diffraction**, since it is diffraction which enables us to construct, or indeed to reconstruct, any specific type of light wavefront.

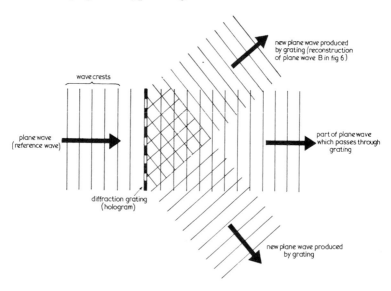

14 Production of plane waves by diffraction. If we consider the grating as a hologram then the plane wave at the top is the reconstruction of Wave B in fig. 6 and the reference beam is the same as Wave A.

Diffraction

When light hits an object which obstructs its passage, the object casts a shadow. However, some of the light is bent; i.e., it does not travel in a straight line—and it bends into the area in shadow. This effect, known as **diffraction**, occurs as a result of the wave nature of light, though its detailed explanation is quite complicated.

Light is bent by only a very small angle into the area of shadow and we do not normally see this. If there are many obstructions, and the distance between them is only a few wavelengths of light, the combined effect of light bending into the regions of the shadows becomes more significant.

When a wavefront hits a single large obstruction the result is merely to 'chop out' a large section of the wavefront. The rest of the wavefront carries on almost unaffected. When a wavefront hits several small obstructions, however, it is so changed by the process of diffraction that, in effect, the light which emerges from the other side is represented by an entirely different wavefront. Its shape will be changed or 'distorted' and it may travel out in an altogether different direction. Diffraction thus provides us with a means with which to convert one wavefront into an altogether different wavefront; i.e., it is a mechanism by which we can construct a wavefront of light.

A **diffraction grating** is a device which creates a new wavefront in this way. The simplest consists of a grating of narrow straight lines, a small fraction of a millimetre apart, on a small glass plate. Plate 3 is a magnified photograph of a rather sophisticated one which consists of fuzzy bright and dark regions rather than well defined straight lines. If we place this in the path of a laser beam, some of the light will pass straight through the grating and some of it will be bent to form two new beams which emerge at an angle to the original beam and on either side of it. If we make the original laser beam a plane wave then the two new beams on either

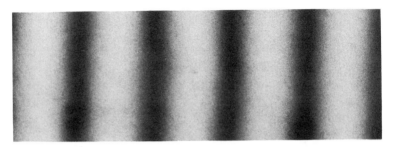

3 Diffraction grating (x 100). Note that this is identical to the interference pattern of Plate 1. In fact this pattern may be regarded as a hologram, recording either of the plane waves in fig. 6.

side are also plane waves (fig. 14). By passing the laser beam through the diffraction grating we have constructed two new plane wavefronts.

You may have recognised the diffraction grating of plate 3. It is exactly the same as the interference pattern described in Chapter 1, produced when two coherent plane waves meet at an angle (plate 1). The diffraction grating is probably the simplest example of a hologram, and we shall now describe its production and use in the terminology of holography.

We start with two coherent plane wavefronts which interfere to produce a pattern which is recorded on a photographic plate by placing the plate at the position of the screen in fig 6. In holography this is known as the **recording** stage of the process. One of the plane waves (for argument's sake, Wave A) is called the **reference wavefront**. Let Wave B be called the **object wavefront** (this is the one which is to be recorded): in this case it is just the same as the reference wavefront, but, when making a hologram of a 3D object, it would be the wavefront of light which is reflected from the object. The interference pattern recorded on the photographic plate (i.e., the diffraction grating pattern) is the hologram. If the hologram is placed in the path of the original reference beam (i.e., a plane wave laser beam) then two new wavefronts accompany the beam when it emerges from the other side (as in fig. 14). One of these is an identical replica of the original object wavefront; i.e., it consists of a plane wavefront travelling in the same direction as Wave B (the other new wavefront produced will be discussed later). This second stage of the holographic process is known as **reconstruction**.

Thus we see that the recorded interference pattern of two coherent plane waves is a device which, when subsequently illuminated by just one of the plane waves, will cause the other plane wave to be reconstructed. The complete process amounts to recording and storing the object wavefront as an interference pattern on the photographic plate (the hologram) and then liberating it at a later time by passing the reference wave through the hologram. In fact, as we shall see, the object wavefront can be any wavefront; e.g., the wavefront reflected from a real object, as long as it is coherent with the reference wave. The interference pattern produced by any two coherent wavefronts is just that device which will later, by diffraction, convert one of the wavefronts into the other. It is the key, first understood by Dennis Gabor, to holography.

The image produced by the simple diffraction-grating hologram may be observed by positioning one's eye about 1m from the grating and looking through the grating in the direction in which the reconstructed plane waves are emerging. Since the waves which the eye is receiving are plane the image perceived is of a plane, which one can think of as a completely blank wall illuminated by light that is the colour of the laser. There is nothing special about it in that plane waves always give you such an image whether they are produced by a hologram or not. Since there is no detail of the 'wall' it would be impossible to tell how far away it is; in fact you would be looking at what was an infinitely large wall an infinite distance away, but you would see only that part of the wall which would be visible through the small 'window' which is the diffraction grating. So, in fact, the diffraction grating would appear to glow uniformly brightly across its surface; not a particularly interesting image to view.

The diffraction-grating hologram demonstrates quite simply a number of effects which are observed with other holograms. The simple grating bends light passing through it into two new beams. Given a particular diffraction grating we may illuminate it with laser beams of colours other than that of the light used to make it. In each case the angle through which the beam is bent depends on the colour of the light. Red light (which has a long wavelength) is bent through a larger angle than blue light (which has a short wavelength). If we pass white light, a mixture of all colours, through the grating, each of the component colours is bent through a different angle and we get a spectrum of colours coming out of the grating like that produced by a prism.

The lines of the grating must be very close together to produce appreciable bending. For example, to bend red light through 20° requires a grating the lines of which are only 0.002mm apart; for finer spacing of the lines the light is bent through a greater angle. To record these lines originally we need a photographic plate capable of recording very fine detail. Also we must ensure that the plate does not move while we are making an exposure and recording the interference pattern. If movement occurs the pattern may become so blurred as to be indistinguishable and we end up with a glass plate which is uniformly grey or black across its surface. This will, of course, not reproduce the special diffraction effects of the diffraction grating.

The type of diffraction grating we have so far considered is known as the **transmission** grating, since it affects light which is transmitted through it. If the lines of the grating were on the surface of a mirror rather than on a glass plate we would have a **reflection** grating. Such a grating would reflect light of different colours at different angles. There are similarly two broad categories of holograms, transmission holograms and reflection holograms, according to whether the image is viewed by the light transmitted through, or reflected from, the hologram.

Transmission Holograms—plane holograms

We are now in a position to describe how to make a hologram of an object, or, more precisely, of the wavefront of light reflected from an object (or objects). An object illuminated by laser light will reflect a wavefront of light which has the all-important property of coherence. It is thus possible for this wavefront to produce a unique interference pattern when combined with another coherent wavefront (usually a simple one such as a plane wave) which acts as a reference beam.

The object is placed in the vicinity of a photographic plate and illuminated with laser light. Part of the reflected wavefront, which travels out in all directions from the object, will fall on the photographic plate. At the same time the reference wavefront (or reference beam) is projected at an angle towards the photographic plate (fig. 15). The necessary coherence of the two wavefronts is assured by deriving both the reference wavefront and the beam which illuminates the object from the same laser using a device known as a **beam splitter**. This produces two separate beams each of which is expanded by a **beam spreader** and directed by mirrors to the

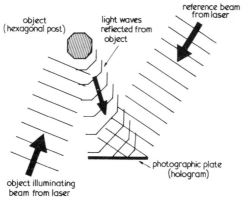

15 Recording a transmission hologram: both the beam illuminating the object and the reference beam are derived from the same laser to ensure the necessary coherence between them.

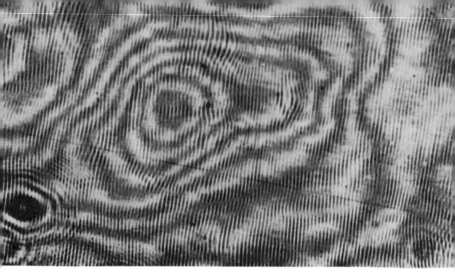

4 Surface of hologram plate, seen in white (incoherent) light (magnified). This complex pattern of dark lines, completely incomprehensible under these viewing conditions, will, when illuminated with coherent (laser) light recreate the actual light waves reflected from an object, producing an image which, like the object itself, is fully 3D.

appropriate positions; one to the plate and the other onto the object (fig. 17).

The photographic plate is exposed to the pattern for a length of time that depends on the brightness of the pattern; i.e., on the power of the laser used. With the comparatively low-power lasers commonly used this may amount to several seconds. The recording process is completed with the development of the photographic plate. The pattern across its surface constitutes the hologram. It consists of a pattern of finely detailed lines the complexity of which bears no apparent resemblance to the original object. Holding the hologram at arms' length in normal daylight, it appears to be almost uniformally grey and gives no hint of the image encoded in its structure. Closer inspection under a magnifying lens reveals an intricate pattern of curved dark lines, whorls and bulls'-eye targets (plate 4). In fact these are just superficial blemishes on the hologram, caused by diffraction of light from particles of dust and defects in the optical system used. The important *holographic* fringes cannot be observed without the aid of a powerful microscope.

The set-up used in reconstruction is the same as that used in recording the hologram except that the object and its illuminating beam are now removed. The hologram is placed in the reference beam at roughly the same angle to it as that used in the recording stage. Some of the beam passes straight through the hologram, unaffected by its presence, but some of it is bent to either side into two other wavefronts, one of which is a replica of the original wavefront reflected from the object. To intercept the reconstructed wavefront we position our eyes about a metre from the hologram in the appropriate direction and look through the hologram (fig. 16). When this wavefront falls on the eyes it produces the sensation that the object is behind the glass plate, at the same position it occupied when the hologram was recorded. The

5 A laser transmission hologram. Under illumination by laser light the hologram acts as a window through which the object (in this case a statue of the Virgin and Child) may be seen apparently sitting behind the plate.

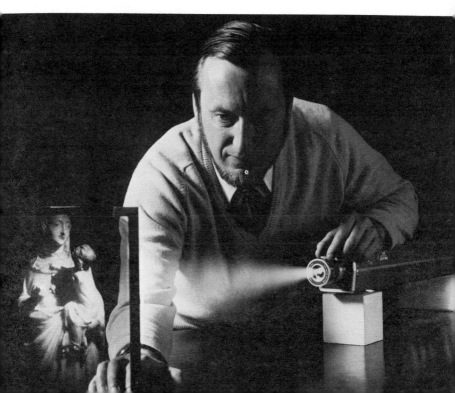

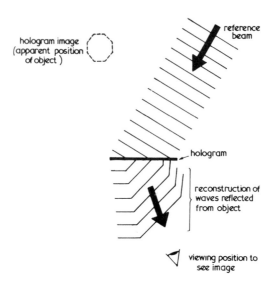

16 Reconstruction of transmission hologram (virtual image). In this configuration the hologram reconstructs the actual light waves reflected from the object when it was recorded. The part of the reference beam which travels straight through the hologram is not shown here for the sake of clarity.

glass plate of the hologram appears as a 'window' onto the scene which existed behind it when it was recorded (plate 5).

The holographic image (as we shall call it) thus produced looks just like the original object. It is three-dimensional and we may look around the foreground objects at those behind them simply by moving our head slightly to one side—the phenomenon of parallax (plate 6). We may see the image only so long as we are looking through the 'window' which is the hologram; thus, if we move our head directly to the side of the hologram we lose sight of the image since it is no longer possible to look at it through the hologram. Of course, care must be taken when viewing the hologram from different angles not to look directly down the laser beam as this could be injurious to the eyes.

A hologram may be cut into two pieces to produce two holograms of the object. However, since each of these holograms is obviously smaller than the original, they each provide us with only a small section of the original 'window' through which to view the image behind. We are therefore restricted in the range of angles from which we can observe the image. Necessarily, the two holograms will produce images of the object from slightly different viewpoints.

The glass photographic plates used to make holograms range in size, exhibition-style holograms being plates up to 1m² in area. The smaller plates range in price, according to size, from several dollars each to several hundred dollars. Besides glass photographic plates it is possible to obtain flexible plastic film which is also suitable for holography. This may be obtained in sheets of up to 6m², allowing very large holograms to be made.

By using flexible photographic film shaped in a cylinder surrounding the object we may produce a **cylindrical holo-gram**. A reference beam has to be produced which will illumi-nate all the inside of the cylinder; a simple plane wave will not do. The result is a hologram of an object which we may walk around and view from any angle. The object will appear to be sitting in the middle of the cylinder.

17 Typical arrangement for producing a transmission hologram.

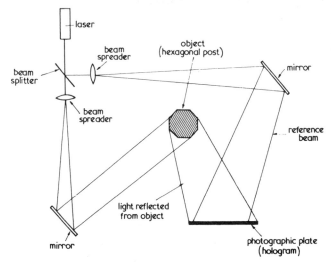

The holographic image has certain characteristics due to the fact that it is made using the highly coherent light of a laser. The object was originally illuminated with laser light of only one colour, often the red light of the helium-neon laser, and the subsequent reconstruction is also made using light of this colour. As a result the holographic image is in this colour, i.e., is monochromatic—appearing to be, for instance, in black and red. Holograms which reproduce the natural colour of objects require a more sophisticated method of production (see Chapter 5). Another characteristic of the hologram image is that it has the peculiar speckled texture which appears when an object is illuminated by coherent light (see Chapter 2). Although attempts have been made to cure holograms of this annoying graininess they have met with only limited success. Holograms are capable of reproducing ranges in brightness of the different parts of an object of over one million whereas for a photograph this figure is about one hundred; thus holograms are extremely good at reproducing the tonal qualities of an object.

To demonstrate the realism in holograms they are sometimes made with pieces of cut glass included in the object scene; as one looks from side to side at the image, light is seen to glint off different facets of the glass—just as it does with the original. Another device is to include a magnifying glass in the scene, placed in front of some small objects; by looking through the magnifying glass in the reconstructed image, the objects behind it appear magnified; as one moves from side to side different objects can be seen in magnification, just as they would were one looking at the original magnifying lens and object.

Two different holographic images may be recorded separately on the same photographic plate by making a double exposure. In the resulting hologram the two images appear superimposed. 'Trick' holograms may be prepared in this way. A first hologram of one object (or objects) is 'exposed' on the plate in the normal way. The object may then be

6 Holographic image from two different viewing positions. Photograph (a) was taken from a point slightly to the left of the position from which photograph (b) was taken. A bolt in the foreground of the scene obscures different parts of the stopwatch face in the two photographs—the effect known as parallax. (This hologram was made by the author using the simple apparatus described in Chapter 8.)

moved slightly or replaced by another one positioned anywhere behind the plate. A second exposure is then made on the same photographic plate. When the resulting hologram is illuminated, reconstruction of both images occurs and we see an image of the two objects superimposed. By positioning the objects appropriately one can create the illusion of two objects overlapping in space, or of one being inside the other.

We said earlier that, on reconstruction, the hologram produces two new wavefronts and that one of these is a replica of the wavefront originally reflected from the object. The image produced by this wavefront is known as the **virtual image** (or true image) and is the one which we have so far considered. The other new wavefront which emerges from the hologram also produces an image, the **real image** (or conjugate image), which has a number of peculiar and very interesting properties. To see this image we must place our eyes in a position in which they may receive its wavefront, as shown in the arrangement of fig. 18, and then look towards the hologram. The real image appears to 'float' in the space between the observer and the hologram; i.e., it is in front of the plate, rather than behind it.

The peculiarity of this image is that it is in a sense a reversed image of the original object (or objects); it appears to be back-to-front, and the normal effects of parallax are also reversed, a phenomenon known as **pseudoscopy** (we call it a pseudoscopic image). Pseudoscopy refers to the effect caused by moving one's head from side to side while viewing the real image; objects in the background of a scene appear to move more than those in front. The effect can be surprising and disconcerting when first encountered since it is contrary to the parallax effects observed in everyday life, known as **orthoscopy**. No description can do justice to the pseudoscopic real image of the hologram; it must be seen to be appreciated. Holographers who have spent some time working with them say that with increasing familiarity these images are converted in 'the mind's eye' to appear as normal orthoscopic images.

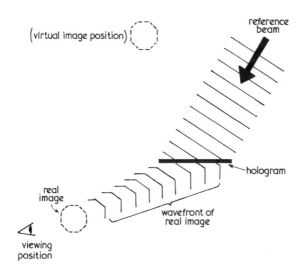

18 Reconstruction of real image from transmission hologram. With this configuration a transmission hologram produces a real image, visible from the position shown and appearing to be in front of the plate. It has the peculiar quality of pseudoscopy (see page 68).

One is reminded by this observation of tests done with human subjects who were provided with spectacles, to be worn at all times, which made the world appear upside down. After an initial week of confusion the subjects reported that they once again saw the world the right way up—their minds had learned to make the transformation. When the test was over, and the spectacles were taken off, they were thrown into another short state of confusion before their minds adapted back to seeing things the normal way up.

It is possible to transform the real image from being pseudoscopic to orthoscopic by a special technique (see Chapter 5).

Another interesting property of the real image is that a screen placed in its vicinity will capture on its surface a flat image, like a photograph, of the object which was holo-grammed. If the screen is replaced by a photographic plate, we may indeed record a photograph of the object by this means, though this could be accomplished far more simply using a conventional camera!

The type of hologram described here is known as a **transmission hologram** since, by transmitting light through it, the wavefront of an object is reconstructed. The technique outlined for its production is known as off-axis holography, to distinguish it from the type of hologram produced in a slightly different way by Dennis Gabor in 1948. The basic principles are the same for both, but off-axis holograms, invented by the US scientists Emmett Leith and Juris Upatnieks at the University of Michigan in the early 1960s, have a number of advantages which make them superior to the early holograms of Gabor.

The versatility of the holographic process is limited in several ways by the power and quality of the laser used in a particular

7 Typical sophisticated holographic apparatus. The subject of the hologram is in this case a flat plate slightly to the left of the centre of the table (it casts a shadow on the laser behind). Note the heavy construction of the table and supporting legs, the guidance of the laser beam by mirrors, beam-spreader lenses and the holder for the photographic plate at the far left of the table. The electronic box is an automatic exposure meter.

set-up and by the conditions required for the recording of the hologram. These limitations apply quite generally, not just to transmission holograms but also to many of the other types of hologram that we have yet to consider.

Most holograms must be made in a darkroom, so that the only light hitting the photographic plate is that of the interference pattern to be recorded.

The spacing between the lines or fringes in the interference pattern of the hologram is about the same as for the diffraction grating (i.e., as many as one or two thousand fringes in each millimetre). Just as with the diffraction grating, high-quality photographic plates, capable of recording this fine detail, must be used in holography. Vibration or movement in the set-up can be a problem too. With the lasers of modest power often used in holography (because of their relatively modest price and good coherence properties), the exposure time for the holograms may be several seconds, since the interference pattern is quite dim. If there is movement in the apparatus or the object by more than roughly a wavelength of light (1/1,000mm!) during the period of the exposure, then the recording on the photographic plate of the interference pattern will become blurred. If this happens the brightness of the reconstructed image is decreased; indeed, if the blurring is very great, the image is lost altogether. For this reason holographers working with low-power lasers often take elaborate precautions to isolate the equipment from vibration and movement. The apparatus being used may be mounted on a heavy granite or metal table which is supported on a firm concrete floor by some kind of shock-absorbing system (plate 7). A shock-absorbing system commonly employed is the supporting of the table on partially inflated rubber tyre inner-tubes; this cuts out practically all vibration. However, in making holograms of some objects, all such efforts are in vain simply because it is impossible to stop the object itself from moving; holograms of people are impossible with low-power lasers since the action of the heart in pumping blood through the blood vessels of the skin causes it to move by an amount greater than that allowable. Another limitation is that we can only hologram objects of a sufficiently small size to fit on the special vibration-free table. (The next chapter will deal with the pulsed lasers—which give out a very bright, very short, flash of light—used to overcome these limitations.)

A limitation placed on holography with the cheaper lasers lies in the coherence length of the laser used. The coherence length of a laser is a measure of the degree of purity in colour of the light emitted from it. Cheaper low-power lasers tend to have shorter coherence lengths; i.e., the light they produce is less pure. In general, when making a hologram the light which is the reference beam and that which illuminates the object will have travelled paths of different length before they meet on the photographic plate to interfere and produce a pattern. For parts of the object close to the plate this path difference will be less than for parts of the object which are far from the plate, since light travelling to, and reflected from, the more distant parts of the object will have to travel further than for light of those parts close to the plate. If the path difference is longer than the coherence length of the laser then the interference pattern will not be produced, and no image may be reconstructed. For lasers of short coherence length (e.g., $\frac{1}{2}$m) this places a limit on the depth of the scene that can be satisfactorily recorded in the hologram. The importance of the laser in the production of holograms of 3D scenes now becomes clear, for while lasers have coherence lengths from roughly $\frac{1}{2}$m to 1km, the sufficiently bright non-laser sources of coherent light rarely have a coherence length of more than a few millimetres. It was only when the laser was introduced that holograms of 3D objects became a real possibility. (In fact, holograms of 2D objects—e.g., transparencies or slides—were possible before the advent of lasers using gas-discharge lamps; Gabor's original holograms depicted a clear plastic protractor.)

When a transmission hologram is illuminated for reconstruction of the image, no such problem of path difference occurs and we may use a less expensive light source with a relatively short coherence length. A gas-discharge lamp (e.g., neon) produces light of a sufficiently pure colour to enable the hologram image to be reconstructed. Indeed, transmission holograms may be viewed in the light of an ordinary slide projector fitted with a piece of coloured plastic in place of a normal slide. Nevertheless, the purer the colour of the light used, the sharper the definition of the image, and so the laser gives the best image.

If the hologram is illuminated by light of a colour other than that used to make it, the image will be in this colour and appear larger or smaller than the original object according to whether the wavelength of the light is shorter or longer than that of the light used to make it. Also, it will appear at a different distance behind the hologram. If the hologram is illuminated with a strong source of white light, a mixture of all colours, then each colour produces a different image and the result is a rainbow 'smear' of images.

Volume Holograms

The transmission holograms we have discussed consist of an interference pattern recorded across the surface of a photographic plate (or film). Photographic plates, as we mentioned in Chapter 1, are glass plates coated with a thin layer of emulsion, a light-sensitive substance. When the plate is developed those parts of the emulsion which have been exposed to light become black and will not transmit light (or, if 'grey', will transmit only part of the light). In fact, the layer of emulsion is somewhat thicker than the distance between the interference fringes in the hologram. It is thus possible under certain circumstances to record the hologram as an interference pattern which extends not only across the surface of the emulsion but, since the emulsion is quite transparent, throughout its depth as well. Such a hologram is known as a

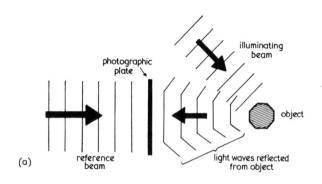

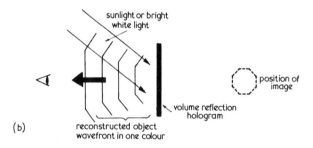

19 The white-light reflection (volume) hologram: (a) recording, using coherent laser light; (b) reconstruction in incoherent sunlight or bright white light.

volume hologram and has a number of properties not shared by the transmission hologram of the previous section (**a plane hologram**). Volume holograms can be made that will produce an image either by light transmitted through them (transmission holograms) or by the light reflected from their surface (reflection holograms).

To obtain efficient reconstruction with a **volume transmission hologram** it is vital that the reference beam used strikes the hologram at exactly the same angle as it did for the recording of the hologram; otherwise no image will be obtained. Also, the light used in reconstructing the hologram must be of the same colour as that used to record it. As a result it is possible to record as many as one hundred separate holograms on the same photographic plate, using a reference beam at a different angle for each one. Each may then be independently viewed by illuminating the hologram at the appropriate angle.

The method used to record a volume transmission hologram is similar to that used for the plane type, the main difference being that one tends to use photographic plates with a particularly thick layer of emulsion on them.

A **volume reflection hologram**, however, is made in an altogether different way. To record one of these we arrange that the reference beam and the wavefront from the object hit the photographic plate from nearly opposite directions (fig. 19a). We may regard the reflection hologram as containing interference fringes which have two separate functions. Throughout the depth of the hologram the fringes act as a colour filter to light which is reflected from the hologram. (This colour filter is of a type known as an interference filter; due to the way in which the hologram is made, fringes in the hologram lie in planes half a wavelength of the laser light apart, each of which reflects some light like a mirror—only light of a colour which has a half wavelength equal to the spacing of these 'mirrors' will have combined reflections which interfere constructively). Of course, the fringes also act as normal holographic fringes on the light reflected by the filter, so that the wavefront from the object is reconstructed in light of this colour. The colour of light which the filter reflects depends on the spacing of the planes of fringes, which in turn depends on the colour of the laser used to make the hologram.

Theoretically, if a red laser is used to make the hologram, the planes of the fringes will act as a red filter, and when white light strikes the hologram it will absorb all the colours in the light except red, which it will reflect. In practice, a reflection hologram made with a red laser often constructs an image in green; this is because of a slight shrinkage in the thickness of the emulsion which occurs when the hologram is processed and which changes the spacing of the planes. Usually the effect of this is of no great concern—since the image is monochromatic it hardly matters whether it appears red or green. It can be a problem when making 'colour' reflection holograms (Chapter 5). An additional effect observed with reflection holograms is a change in the colour of the image as it is viewed from different angles; e.g., a change from green to blue.

Thus a volume reflection hologram will produce a 3D holographic image even when it is illuminated with white light. The best results are obtained by using a source such as the sun or a bright spotlight, and reflecting the light from the hologram's surface (fig. 19b). These holograms are often called 'white light reflection holograms' or 'Denisyuk holograms' after the Russian who invented them (1961).

Phase Holograms

Holograms may be recorded in media other than photographic plates. There are a variety of materials sensitive to light and capable of resolving the fine detail required. These other materials do not respond to light in the same way as a photographic plate. Instead of being darkened by light hitting them at a certain point, their thickness at that point is altered. Light waves which subsequently pass through this point will be changed in phase according to the thickness of the material there. On reconstruction, the phase of the reference wavefront is altered point by point across the surface of the hologram; on emerging it replicates the object wavefront as required.

Such **phase holograms** can be made either of the transmission or of the reflection type and produce brighter images than photographic plate holograms, since more of the light which hits them is bent to form the reconstructed wavefront. Phase transmission holograms viewed in ordinary daylight appear completely transparent, like a sheet of glass.

One such medium for recording phase holograms is a special clear plastic, known as thermoplastic, used in the form of a film placed in the position of the photographic plate in fig. 15. To record holograms in thermoplastic generally requires high-power lasers or long exposure times since the hologram is recorded by the heat of the interference pattern melting the plastic slightly and causing it to distort in the appropriate way. Another phase medium commonly used is dichromatic gelatin. (Plate 18 is a photograph of a holographic pendant made using dichromatic gelatin.)

Phase holograms on plastic film have become important because it is possible to reproduce them — even 'mass-produce' them — without light. A 'master copy' is made from the original hologram on a metal plate, by a process called 'photo-etching'—literally, 'etching with light.' This hologram has the same microscopic pattern of thickness as the original hologram, but it is strong enough to stamp that pattern into sheets of plastic film, under pressure and heat

8 Good quality phase hologram. This photograph, of an image produced by a bleached phase hologram, is practically indistinguishable from a photograph of the actual objects—demonstrating the high quality of the images which may be produced.

in a special press. This holographic printing or 'embossing' can be completely automatic, and very quick. Small plastic holograms are now being reproduced for less than ten cents each. These have been used as novelty badges and giveaways. They have even been 'printed' onto credit cards to make the credit cards harder to forge.

In fact, ordinary photographic plate holograms may be converted into phase holograms by bleaching the dark fringes using a suitable bleaching agent. The variations in darkness across the pattern on the plate are then converted into varying thicknesses to alter the phase of the reference wavefront. This combines the convenience of the photographic holograms (low-power lasers can be used) with the brightness of the image obtained from the phase holograms (Plate 8).

5 Display Holography

The fascinating 3D images produced by Leith and Upatnieks in the early 1960s stimulated many research groups to enter the field of holography. Besides improving the technique of producing pictorial holograms, known as 'display holograms', much of the research was aimed at using holography in a number of ways in which the images produced have scientific and technical uses. It was hoped that holography might emerge as a 'new technology' which could be applied to the solution of a number of very different problems.

Display holography makes possible the display of 3D images in an entirely new way. It thus qualifies as a visual 'medium' which may be useful in a wide variety of applications. We shall consider now the technical developments in display holography, and then go on to examine the potential of holography as a medium and some of the ways in which it has already been used.

Technical Developments

The original off-axis transmission holograms of Leith and Upatnieks had such strict conditions for their production and viewing that, for some time, holography was confined to the laboratory. Only those objects which were small and remained absolutely still during an exposure of perhaps minutes could be 'holographed'. Thus it was impossible to make holograms of people. The images were monochrome; that is, the image was of an object illuminated with, say, red light so that they did not have the same colour as the original object viewed in daylight. These first holograms depicting 3D objects required a laser to make them and a laser, or at least a bright source of monochromatic light, to view them.

Developments in display holography have been concerned with improving the quality of the images, producing full-

9 Holographic portrait of Dennis Gabor. The real Dennis Gabor, inventor of holography (in 1948), stands to the right of his holographic doppelgänger, seen seated at a desk. The hologram was made, using a pulsed ruby laser, shortly before Gabor was awarded the Nobel Prize in 1971 in acknowledgement of his invention.

coloured images, allowing moving objects to be holographed, producing more exciting images and reducing the necessity of coherent sources to view holograms. With regard to the last, the white light reflection hologram demonstrated in the mid-'sixties was the first major step. Since then there have been others; developments in the techniques of display holography have allowed it to emerge from the laboratory.

Holograms of People

As we mentioned in Chapter 4, it is very important in making a hologram that an object does not move while the exposure is being made. Stability is usually obtained by mounting all the apparatus involved, including the object, on a large heavy table which is supported by some sort of vibration-damping system; but this limits the size of the object which may be hologrammed since it must be small enough to fit on the table.

Another way around the problem of object movement is to decrease the exposure time by using a pulsed laser, which gives out a powerful flash of light lasting only one thousandth of one millionth of a second. This enables objects which are moving at a speed of several metres per second to be effectively 'frozen' for the period of the exposure; for example a hologram can be made of a bullet in flight (plate 10). Using a pulsed laser, a vibration-free table is no longer necessary, so that it is possible to produce holograms of large or immobile objects as long as one ensures that the object and the holographic apparatus, including the photographic plate, are in darkness (except for the appropriate laser illumination) for the period of the exposure. This situation may be obtained by placing the apparatus in a darkroom or, conceivably, if the object is immobile and out of doors, by making the hologram at night. Often pulsed lasers have relatively short coherence lengths, and this can place limitations on the depth of the scene to be recorded in the hologram (see Chapter 4). Pulsed lasers are generally expensive and not freely available.

Producing holograms of people presents several problems, the main one being that during an exposure of a hologram lasting several seconds the subject will certainly move by more than half a wavelength. Accordingly pulsed lasers must be used. When making a holographic portrait with a pulsed laser care must be taken to minimise the radiation which hits the eyes of the subject, since it can be very dangerous. This is done by diffusing the light used to illuminate the subject by passing it through a large sheet of frosted glass. In the USA it is now possible, at considerable expense, to commission a portrait by a holographer. (Plate 9 shows a holographic portrait of Dennis Gabor, inventor of holography.)

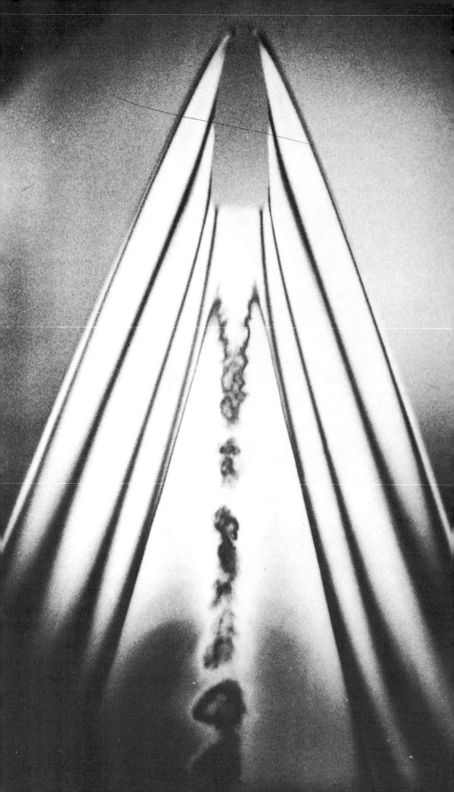

Holograms in colour

The aim of colour holography is to produce a 3D image of an object which has the same colour as the object itself when seen in daylight. Any colour can be produced by mixing appropriate proportions of the primary colours red, blue and green. Were it possible to make three holograms of an object, using red, blue and green laser lights, and to superimpose these images, then the result would be an image which was the same colour as the object. There are, however, a number of practical problems involved in any attempt to achieve this.

The simplest way of superimposing the three hologram images is to record each of the holograms on the same photographic plate. Subsequent illumination of this combined hologram simultaneously by red, green and blue lasers should give a full-colour image. Unfortunately, things are not that simple. In the case of a plane transmission hologram, the green hologram, for instance, will work as a hologram for blue and red light as well, producing two extra images, one larger and one smaller than the original object. The same is true for the other two holograms, and so the resulting product is one full-colour image and a number of other images in the wrong position and of the wrong size and colour.

The simplest solution is to use a white light reflection hologram. Reflection holograms made with red, green and blue laser light may be recorded on the same photographic plate. Each of these holograms will have its own interference filter which will reflect light to create an image in the appropriate colour; each thus produces an image in a single colour, and all are superimposed. The spurious images obtained in the case of the transmission hologram are lost. The problem with this system is that the shrinkage in the emulsion (discussed in Chapter 4) changes the colours for which the three

10 Made using a pulsed laser by a technique known as double-exposure holographic interferometry (Chapter 6), this hologram shows the shock wave accompanying a bullet in flight.

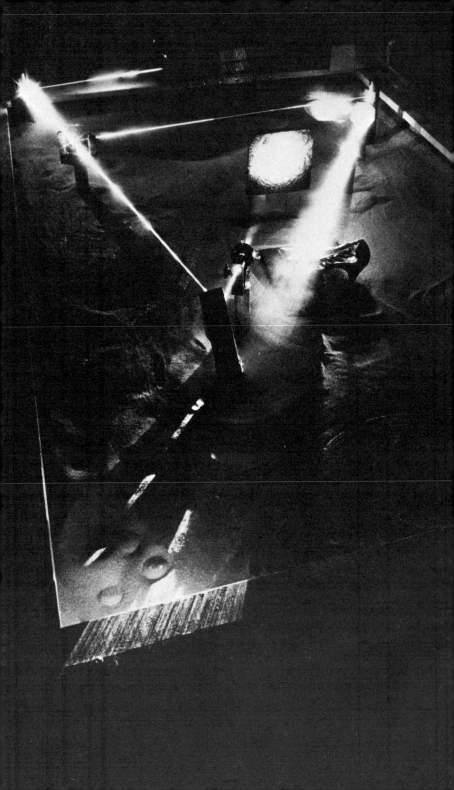

interference filters work. This in turn distorts the colour of the reconstructed image.

In order to maintain the correct balance of colours to recreate an image faithful in colour to the original object, it is necessary to devise some means of minimising this shrinkage. This may be done by, for instance, pre-swelling the emulsion, and reflection holograms have been made which recreate coloured images with convincing accuracy. Perhaps their biggest advantage is that, like any reflection hologram, they may be viewed in the white light produced by a small torch or spotlight and do not require three different coloured lasers to reconstruct the images.

'Stand out' images: projection holograms

A transmission hologram produces two images: a 'virtual' one which appears like the original object behind the glass plate and is orthoscopic; and a 'real' one in front of the plate which has reversed parallax and so is pseudoscopic (see page 68). Either of these images is visible when the eyes intercept their wavefronts, which emerge in different directions from the hologram. Using an ingenious technique it is possible to make a hologram which produces a real image that has the correct parallax and appears to stand out in front of the plate. When viewing such a hologram, it seems that an object with normal parallax is simply floating in the space in front of the glass plate.

This is the nearest that we may get to projecting a holographic image into space; it has one important limitation—it may not be viewed from all angles since the glass plate must always lie behind the image for it to be seen. Such holograms

11 Simple setup for recording a two-beam laser transmission hologram of an object (a shoe). The laser (top) is directed by small mirrors, split into two beams expanded by lenses, one lighting the subject, the other becoming the reference beam and directed to the plateholder facing the subject (photo credit: Mike Sage).

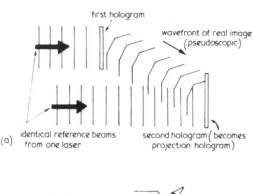

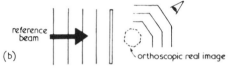

20 The projection hologram, which produces a 3D image which appears to stand out in front of the hologram plate: (a) recording a projection hologram; (b) reconstruction of a projection hologram.

are often called **projection holograms**.

The trick is to make a hologram of a hologram. As stated in Chapter 4, holography can be used to record any coherent wavefront. There is no reason why we should not be able, by making a second hologram, to record the real image wavefront produced by a first hologram. (An arrangement for doing this is shown in fig. 20a.) The second hologram produces a real and a virtual image, but now the real image is the orthoscopic (normal) one and the virtual image is the pseudoscopic (reversed-parallax) one; so the types of parallax observed in the two images have been exchanged. The virtual image now has the peculiar reversal of parallax. By moving our head to observe the real image we see that it is now the right way around and has normal parallax: it appears to stand out between us and the plate (fig. 20b). Just how far in front of the plate the image may be 'projected' depends on the size of the plate used to make the hologram. With a hologram which is 60x90cm the image may extend at least 45cm from the plate—with a 25x25cm hologram it could extend only about 20cm from the plate. Some people have difficulty in focussing on this image since it is rather unexpected, and their eyes tend to try to focus on the plate behind.

The second hologram may be a transmission hologram or, by a somewhat similar technique, a white light reflection hologram.

This type of image is perhaps the most exciting and dramatic of holographic images which can be produced. When viewing such a hologram one is compelled to reach out and touch the apparently solid object floating in space, only to discover, in so doing, that it is completely intangible.

Rainbow Cylinder Holograms

The advantages of white light reflection holograms make them particularly useful for display holography. They may be viewed using simply a white penlight or spotlight and, for this reason, they are a popular medium for holographic display purposes. In 1969, while working for the Polaroid Corporation, Stephen Benton demonstrated a new technique which allowed transmission holograms to be made which could be illuminated with white light (Plate 11). What is more, these new holograms used nearly all the light which hit them to reconstruct the (object's) wavefront and consequently produced very bright images.

The holograms are made by a complicated two-stage process in which a hologram is made of a hologram (Leith 1976). The resulting hologram acts as a prism, bending white light which hits it into different vertical angles according to the component colours. When this hologram, known as a rainbow or spectral hologram, is illuminated by transmission of white light, the observer sees a 3D image behind the hologram, illuminated in a particular colour according to what height it is viewed from (fig. 21a); none of these colours is, in fact, related to the actual colour of the original object. In recently developed versions of this type of hologram the change in colour is not obtained—the image appears in black and white, no matter what height it is viewed from.

Because of the way the hologram is made, the vertical parallax is lost: as we move our head up and down we always see the same view of the image, although the colour changes. We still have horizontal parallax: we can look around foreground objects by moving our head to one side, and this is all that is really important in most cases, since in everyday life it is horizontal parallax that we use most often. Indeed, most people who view such a hologram are not immediately aware that vertical parallax has been lost.

Rainbow holograms have been combined with a technique known as **composite** or **multiplex holography** to produce cylindrical holograms that may be illuminated simply by a bright white lamp placed beneath the cylinder. Again there is a colour shift according to the height at which the hologram is viewed, and a lack of vertical parallax, but the result is a bright image apparently in the centre of the cylinder, which may be viewed at any position around the cylinder.

These composite (or multiplexed) holograms are quite different from the ones so far considered. They are in a sense 'synthetic', working on the principle used in stereo viewers (see Chapter 1) of creating the illusion of a 3D image by sending to each eye a 2D photograph of the object taken from a different angle. Unlike the stereo viewer, they are not limited to one point of view; the composite hologram stores the appropriate stereo pairs taken from all directions.

Looking closely at a composite hologram reveals that the cylindrical film with which it is made is divided into thin vertical strips 1mm wide and 20cm long. Each of these is a separate hologram made independently of the other strip holograms. The subject of each of these holograms is a photograph of the object taken from the position of the strip. When viewing the final composite hologram the left eye and the right eye look through different parts of the cylindrical film (fig. 21b). Each sees a different photograph produced by the

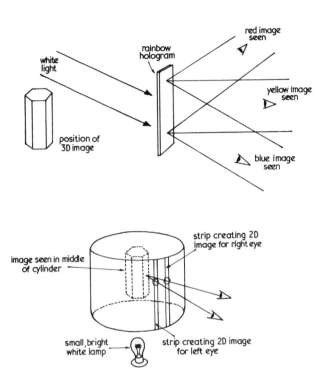

21 These rainbow holograms produce bright images when illuminated by transmission of white light: (a) coloured images produced by rainbow hologram; (b) the rainbow cylinder hologram (composite or multiplex), demonstrating how it produces the effect of seeing a 3D object at the centre of the cylinder.

two different strip holograms through which they look. Each eye thus sees a photograph of the object taken from a different angle (just as with the stereo viewer) and, as a result, we have the impression of seeing the object in 3D. This occurs from whatever point we look through the hologram and gives us a 3D view of the object appropriate to the place from which we view it. One can observe horizontal parallax by moving one's head from side to side. As we walk around the hologram we always see a 3D image in the centre of the cylinder. There are, of course, slight 'jumps' in parallax as we move from one strip pair to the next but these are so small as to be practically imperceptible.

The collection of photographs of the object from different angles, necessary to make this type of hologram, may be made quite simply with a cine camera. Either the object is placed on a revolving table which turns slowly while the cine camera films it, or it is kept stationary while the cine camera goes round it. Using the processed film we can make a series of photographs to be hologrammed. Even more simply, we can mount individual frames from the film in a kind of laser slide-projector and project the pictures onto a screen. Since a laser, which is coherent, has been used, the projected pictures may be hologrammed immediately. To make each of the individual strip holograms, the rest of the hologram film must be masked off and a reference beam provided at the strip to interfere with the light from the projected picture and form the hologram. (In fact, when the multiplex technique is combined with Benton's rainbow hologram technique, as is usually the case, the whole process is rather more complicated than this.)

Since we need only make a movie film of an object to produce a hologram of it we obviate some of the limitations of conventional holography. The subject does not have to be illuminated by laser light, the original movie film being made in daylight or artificial floodlight. Thus it is possible to make a hologram of an outdoor scene. We no longer have the same stringent conditions prohibiting any movement of the object. Composite holograms can be made (quite simply) of people, providing they stay reasonably still during the forty-five seconds or so that it takes to travel around them with a movie camera. Indeed, if the subject makes a simple movement, such as blowing a kiss, while the camera is making its circular tour, this may be translated onto the composite hologram to create a simple moving hologram. As one walks around the holo-gram, retracing the path of the movie camera (or as the hologram is rotated), the subject is seen to re-enact the move-ment made.

12 Rainbow multiplexed cylinder hologram, illuminated by a bright white bulb placed beneath the cylinder. This particular hologram was used in making the film *Logan's Run* (MGM) and depicts the actor Michael York.

The images in composite holograms, unlike conventional ones, do not have to be life-size. The object will appear the same size as the picture projected on the screen and we may alter the size of this by moving the projector towards or away from the screen in the normal way.

Although the projected pictures are in black and white, the final image will be given colour by the prism action of the rainbow hologram, according to what height it is viewed from. The hologram is illuminated by a small bulb placed beneath the centre of the cylinder and may be held simply in a cardboard frame which keeps the flexible hologram film in the correct cylindrical shape. The images are bright and clear but generally speaking of poorer quality than those of conventional laser-illuminated transmission holograms (Plate 12).

The type of cylinder hologram most commonly seen is not a full 360° cylinder, but a 120° 'segment' of this cylinder, mounted in a cardboard holder. It works in exactly the same way as the full cylinder hologram.

The rainbow cylinder hologram, because of its wide applicability and simple viewing, has enjoyed much popularity in display holography. Companies exist that produce custom-made rainbow cylinder holograms. Once made, the holograms may be easily and cheaply copied.

Holograms made by Computer

The object of making holograms with computers is to create a 3D image of an object which has no real physical existence. This may be very useful in design and architecture since it enables a designer to get the feel of how the design would look when actually made. Automobile designers can present their new idea for a car and architects can give their client an idea of how a building would look on completion. Artists could use computer-generated holograms to produce 3D images of imaginary objects or abstract patterns. A more straightforward use would be the display of information resulting from calculations the computer had made, in the form of 3D graphs. It has been suggested that such a system could be used to display the positions and paths of aircraft, enabling flight controllers to be immediately aware of any impending mid-air collisions.

The simplest and most effective method of making holograms with computers is to use the technique of composite holography described in the previous section (page 88). The computer is then programmed with all the necessary 3D information about the object's shape, and instructed to make a series of 2D drawings of the object as seen from all the points on a circle around it. These drawings then become the subjects for a composite hologram of the object, giving an image in 3D. Since only line drawings have been used, the finished hologram will be a 3D 'line image' of the object showing only the edges of the object and no surface detail.

Attempts have also been made to use computers to calculate

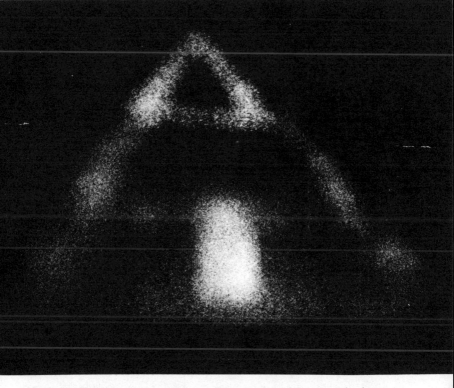

13 Computer-generated hologram. This pyramid has and had no real physical existence; it is a reconstruction from a hologram, the actual form of which was calculated and drawn out by a computer.

the shape and form of the actual interference pattern that would be produced in the hologram of an imaginary object. The computer draws the hologram on a large scale; it is then reduced by a photographic technique to the correct size and observed under coherent illumination. Unfortunately the images produced so far have been rather crude (plate 13). The main problem is that the calculation of the pattern of the hologram is a very lengthy one, too long even for a computer. This is because a hologram contains a very large amount of stored information.

Holography as a New Image Medium

With the invention of holography a new medium was created for the display of visual images. Holography has often been described as 'three-dimensional photography' although, as we have seen, the physical principles upon which it is based are fundamentally different from those of conventional photography. In considering the potential of holography as an image medium it is useful to compare it with photography, yet each of these media possesses unique and distinguishing characteristics. Occasionally, when considering some particular aspect of holography, it may be more appropriate to compare it with other established media—for example, sculpture.

At the time of holography's conception in 1948 photography was already a well established art, both creatively and technically. At times, it has been asserted that photography provides such a faithful reproduction of nature that it leaves no room for creativity and personal expression. In common with many others, I would disagree with this appraisal, arguing that there are several ways in which photography suffers important failures in its attempts to represent reality, and that it is in these so called 'aberrations' of the medium that the photographer may find space for creative expression and the valid communication of ideas. For instance, a photograph is a view of the world from one particular vantage point, and this vantage point is chosen by the individual photographer. The photographer is saying to us: 'Look at this collection of objects (or this scene) from this viewing position'; thus discriminating against all other views of these objects. It is a property of the medium which is usually taken for granted, and its importance and usefulness lie in the fact that it *is* so easily taken for granted because photography actually parallels the way we ourselves perceive the world, always from one particular viewing position. (Another 'aberration' of photography which may be creatively exploited in a similar way is 'depth of focus', by which the photographer discriminates against objects which are outside a certain range by making their images appear out of focus.)

Another important property of photography, which we mentioned in Chapter 1, is that it produces an essentially flat, two-dimensional representation of the world incorporating the notion of perspective. That these representations should be intelligible images is also a fact which is more or less taken for granted.

Although there are other aberrations and properties of photography which may be subject to creative exploitation, the formation of a flat image and the discriminating power of the photographer are the features which I consider most important in a comparison with holography.

In a hologram the images are of course three-dimensional but what is the significance of this property in terms of the usefulness of the medium? When one is first confronted with

14 Holography as a means of cataloguing. This portrait of a mummy's skull was made as part of a programme at the US Smithsonian Institute for the purposes of providing 3D copies of valuable artefacts in danger of decay. The hologram is a white-light transmission hologram which does not suffer colour changes as it is viewed from varying heights.

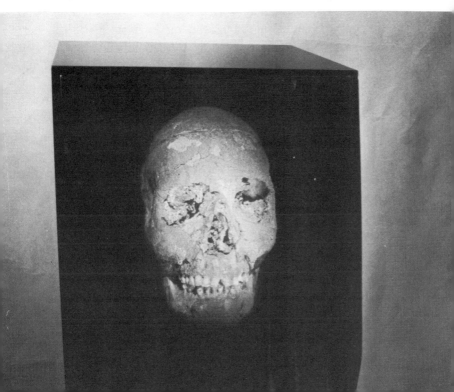

a simple hologram, it is the sheer novelty of seeing apparently solid objects which float in space and yet have no real physical existence which is the source of interest and entertainment. After one has seen half a dozen of these simple holograms, the excitement begins to pall and one considers more deeply whether straightforward still-life holograms of objects are really more interesting than looking at the real objects themselves, which can be considered as sculpture. If holography is to be successful as a medium, enjoying popular use in applications by artists and others concerned with image display, it must possess unique characteristics which give it a special appeal for certain applications.

Holography, in common with photography, provides us with a visual replica (i.e., image) of an object which is faithful in almost all respects to the original. If we are concerned with forming a replica of an object for demonstration or cataloguing purposes, then holography has an obvious advantage over photography, since its replica is three-dimensional and thus viewable from any angle (plate 14). It has been suggested that holography could be used to compile libraries of the best sculpture from all over the world; large audiences would then be able to see these pieces in all their three-dimensional grandeur. Similarly, investigations have been made into the use of holography to preserve copies of old statues which are in danger of decay. Replication, however, is not confined simply to these dry cataloguing activities; it offers real chances for creative expression with such subjects as portraiture and landscape, and in these cases the 3D property of the holographic image is obviously an important advantage.

An important distinction between holography and photography is in the power of discrimination which the artist can exercise in determining the viewer's vantage point on a scene. In displaying a pictorial hologram one is inviting the observer to look through a window at a collection of objects, composed in a scene, and leaving it to the viewer's discretion to examine it from the various possible vantage points. (Obviously this is less true if the scene is placed a long way behind the hologram plate, effectively limiting the angles through which it may be viewed—in doing this much of the 3D effect will be lost anyway.) There is an important type of creative photography which relies for its success on the ability to direct the observer's attention to some detail or a particular viewpoint; it is a class of 'image-making' from which holography, by its very nature, is barred.

Those who use the medium, particularly artists, must be aware of these 'facts of life' about its capabilities. Holography does have unique properties, and even its own special aberrations, which are, arguably, capable of creative exploitation; e.g., double-exposure techniques, interferometry techniques (to be considered in the next chapter) and the ability to form holograms of abstract 3D light patterns. Too often holographic images are expected to impress the viewer simply by their novelty and unexpectedness, and this is often coupled with descriptions in which holography is portrayed as a piece of science fiction made fact, cloaked in an aura of fantasy and mystique.

Compared with other media, the potential of holography remains relatively unexplored. The main reasons for this are the expense of the holographic apparatus and the technical expertise that is required to make holograms. A top-quality holographic studio costs thousands of dollars and requires skilled technicians to manipulate the equipment and process the holograms. Attempts at reducing the cost of the finished product by mass-production techniques have so far had only limited success. Because of the expense there are only a few individuals who can claim to be 'holographic artists', and they are usually affiliated to private or university laboratories. The cost and technical complexities of holography make it difficult for existing artists to understand and use the medium effectively. Thus holography, along with other forms of technological and conceptual art, forces a redefinition, or at least a widening, of the term 'artist' to include

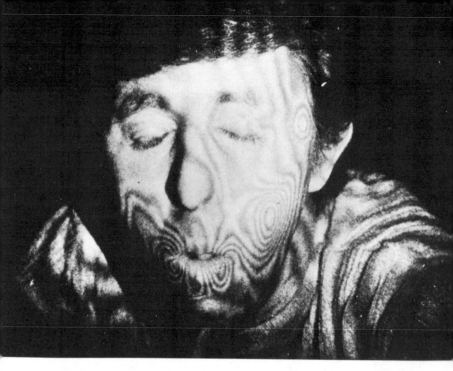

15 Double-exposure holographic interferometry portrait of a man's face. The process that produces the pattern of 'tattooing' across the face of the man is a special property or 'aberration' of the holographic medium.

individuals who not only possess some kind of 'aesthetic sense' but who are also scientifically literate. Technical limitations—such as the size of present holograms and the need, often, to record them in a darkroom—can place a frustrating restraint on the artist's imagination. Many good ideas must be scrapped simply because they are practical impossibilities.

As holography develops as a medium, incorporating some of the possible innovations we shall discuss in Chapter 7, we can expect to see many new and exciting applications which make use of its special characteristics. With present works of art it is difficult enough to define the nature of the interaction between the observer and the art object, but with holograms of the future it may become even more so. Although with present holograms informed observers are aware that they are viewing a holographic image with no real physical existence (due to the presence of such phenomena as speckle in the image), there may come a day when the quality the image is so good that the sense of touch is the only reliable means of distinguishing holographic images from real objects. We would then have the interesting situation in which the medium itself effectively disappears; i.e., the observer is no longer aware of its existence. Moving computer-generated holograms may be made in the future which respond in some way to the presence of the observer (as do some present-day works of technological art). If this stage is ever reached, then it is certain that psychological considerations must be taken into account to give a valid description of the work of art.

Jonathan Benthall, in his book *Science and Technology in Art Today* (Thames and Hudson, 1972), says, in a chapter devoted to holography:

> If holography is as radically a new medium as is suggested, it will develop not only in ways that are predictable, but also in new and unpredictable ways. It will, over the years, influence our art, our every day perception, our language, our reality.

We shall now consider some of the first tentative steps that have been made in using this new medium. The uses of a medium such as photography or painting are often divided into the so called 'fine art' applications and the commercial applications, or 'graphic arts'. While not being an entirely satisfactory distinction, it will be a convenient one for us to make in discussing how display holography has been used.

Use of holography in fine art

There are several artists who have worked with holography exploring its peculiar characteristics and its 'aberrations'. I shall not be able to give a complete and authoritative review of work which has been done in the medium but I hope to show, in describing the work of just a few artists, how holography can be used and to indicate some of the possibilities.

Margaret Benyon, who had previously taught painting, was one of the first artists to turn to holography; before taking up holography she had experimented with a technique for producing 3D stereoscopic paintings which were viewed by wearing a special pair of spectacles. Between 1968 and 1971, while holding a fellowship in Fine Art at Nottingham University, UK, she was engaged in making holograms in the university's engineering department, on a number of occasions in collaboration with engineers from the British Aircraft Corporation. In 1970 these formed the first London exhibition of holography. Margaret Benyon has stated that her holograms deal with phenomena peculiar to the medium. In a double-exposure hologram she has made objects appear to float in space, and a milk bottle can be seen by looking through an orange. In

16 *Interference Pattern Box*, 1969. A transmission hologram, reconstructed by laser, made by Margaret Benyon in collaboration with Peter Spicer of BAC. The frame of the hologram is made from the same wood as the box which is the subject of the hologram.

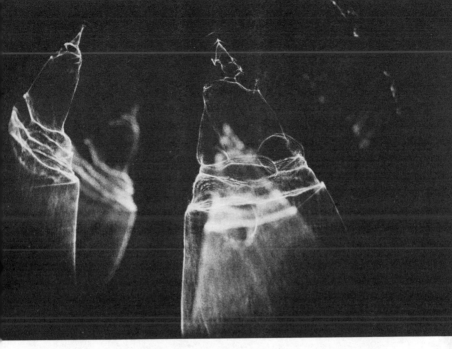

17 (a) *Holos 17*, an abstract light pattern by Harriet Casdin-Silver in collaboration with Stephen Benton (1973). This is a laser-illuminated hologram, the real image of which is projected into the space between the viewer and the hologram plate. When seen in real life the abstract light patterns are shaped like the skins of the segments of an orange.

her 'Hot Air' hologram convection currents rising from a cup of tea during the original exposure are made visible as dark wavy lines, and a hand held in the centre of the scene is left unrecorded (by virtue of its movement); in the final hologram the volume of space occupied by the hand appears as a solid, black, hand-shaped void—a kind of three dimensional 'silhouette'.

Harriet Casdin-Silver is another artist who has investigated the unique properties of holography. She has worked in collaboration with physicist Stephen Benton (inventor of the 'rainbow' white light transmission hologram) producing both conventional holograms, of the laser transmission and reflection types, and also the more recent white light transmission holograms. Although her early holograms were more or less straight holograms of objects (plate 17b), she has taken a step away from 'holographic sculpture' by producing

101

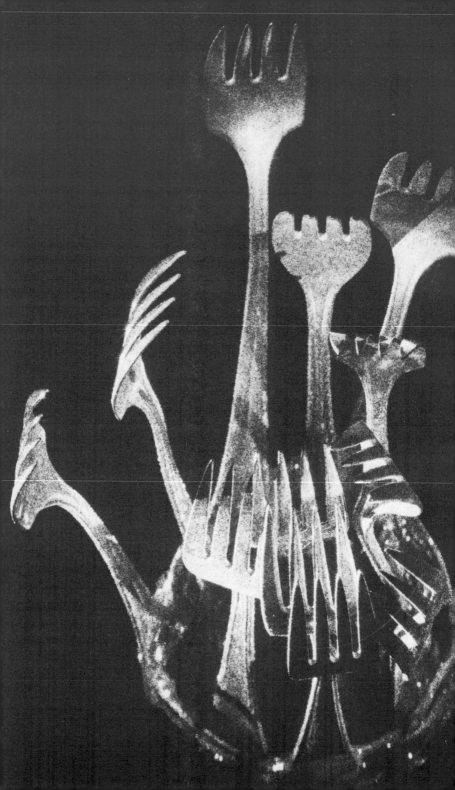

holograms using abstract 3D patterns of light. Casdin-Silver is an assistant professor of physics (research) at Brown University in the USA, and a fellow at the MIT Centre for Advanced Visual Studies. Many exhibitions of her work have appeared in the USA and she has had much experience in using the medium. She has said that it is not true 'that no art has been created in holography. Though there may be little of it, to me the problem is more one of educating the viewer or participant to the unique nature of holographic integrity.'

Salvador Dali has produced a number of holograms—or, at least, provided holographic technicians with ideas for holograms. In 1973 he produced a holographic portrait of rock star Alice Cooper called 'Popstar'. At one time, New York's Knoedler Gallery was asking $25,000 (£12,000) for a Dali hologram original.

Carl Frederik Reuterswald is a Swedish artist who has worked with both lasers and holography. An exhibition of his holograms in Los Angeles in 1971 included an 'autohologram' —a hologram of the laser which was used to make the hologram. In the programme notes to one of his exhibitions he presents 'Hologram without Laser':

> While you are reading these words and displacements are occurring in your consciousness (we hope), mutations are also happening in the space around you and the space within you. In the same way as environment and consciousness exchange action and thought, the body (yours) and space (ours) literally change volume.

A number of museums, such as the Buffalo Museum of Science, USA, and the Science Museum in London, have permanent exhibitions of holograms. In 1976 the New York

17 (b) *Equivocal Forks I*, transmission projection hologram by Harriet Casdin-Silver in collaboration with Stephen Benton (1977). The 3D image of these forks is projected in front of the hologram—compelling the viewer to reach out to try to touch that which has no physical existence.

Museum of Holography was opened to provide a venue for exhibitions of holography and to encourage developments in the medium. The museum holds half a dozen exhibitions annually of different types of hologram and of works by various artists in the field. It has a bookshop which sells textbooks on the subject and a range of holographic products, holograms and jewellery (pendants). It also sponsors travelling exhibitions and coordinates courses on holography.

Commercial holography: Graphics

Holography has attracted considerable commercial speculation, not all of it successful. Many of the early companies that set up in the 1970s to exploit the commercial potential of pictorial holography were small-scale, low-budget operations nick-named 'basement holographers' because of the sites where they often chose to work. A typical survivor of that era is the Dikrotek International Corporation, of Draper, Utah, now a public company. Under the original name Electric Umbrella Limited, company president Richard Rallinson set up 'basement' production in the early 1970s of 'junk jewellery' hologram pendants, using a disc-shaped white-light reflection hologram recorded on a material called dichromated gelatin. More than three million of these gold-coloured pendants have been sold by the company, depicting such subjects as a three-dimensional crucifix, signs of the zodiac, and the pyramid of Cheops. Now several US companies manufacture such pendants. Like Dikrotek, many of these other companies also now make dichromate holograms for industrial uses, not just for jewellery.

Cheap holograms 'embossed' on reflective-silver plastic film are now sold alongside ordinary postcards and posters in many bookstores and novelty shops. Some artists have used embossing to publish cheap 'unlimited editions' of their work. Embossed holograms are also being used on badges, as advertising giveaways and on credit cards. Plastic holograms are making the first 'holographic signs' possible—incorporating many vivid spectral colours as multiply-exposed rainbow holograms, these are set to create a modern-day counterpart to the traditional neon advertising sign in store windows and sales counters. One of the first such signs was made by Global Images of New York for the Tiffen Manufacturing Corp: it advertises Tiffen's special-effect camera filters in camera shops and shows the company's name in brightly coloured lettering along with an image of a camera whose lens filter changes colour as you move.

In September 1982, the English rock band 'UB40' released what it claimed was the world's first-ever holographic record cover, a plastic hologram image of the band's

18 Hologram pendant. A white-light reflection hologram recorded in dichromatic gelatin, the pendant produces an extremely bright 3D image in ordinary daylight.

name in solid-looking 'carved-rock' lettering. The video game manufacturer Atari has been developing experimental electronic games incorporating plastic holograms, but only as scenery so far; more interesting is the eventual prospect of video games where the action takes place in all three dimensions.

Advertising is a good potential sponsor of work in holography; as early as 1972, the New York shop of Cartier's, the international jewellers, displayed a hologram in its Fifth Avenue window. Passersby saw a woman's hand apparently reaching out of the window and holding a diamond-encrusted bracelet over the sidewalk. If anyone reached to take the bracelet, their hand passed straight through. This hologram was produced by a special holography group inside the Conductron subsidiary of the McDonnell-Douglas Company. Although marginally profitable, the parent company eventually closed the holography laboratory, apparently because its activities were just too different from the organization's traditional line of aerospace activity.

Despite the costs, some sort of market has been established for this Cartier-style of custom-made holographic spectacular, using photographic plates up to one meter square in area, or sheet film in even larger sizes. The Japanese car manufacturer Mitsubishi 'projected' its diamond-shaped logo from two one-meter-square holograms at the entrance to its stand at the 1979 Tokyo Motor Show; the holograms were made by the US firm Cambridge Stenographics. In October 1981, crowds gathered outside the Paris department store Galeries Lafayette to see what appeared to be the mask of a woman's face floating above the sidewalk; the image was created by laser-lit holograms in the store's windows. By the early 1980s several European companies had set up to make holograms for these extravaganzas: Holo-Laser and Ap-Holographie in France, and Advanced Holographics of London. But the world's largest holograms—two meters wide and one meter high

—were made for the latest Disneyland development in Florida, the city of the future called 'Epcot' (for Environmental Prototype Community of Tomorrow). One of these holograms shows an aquarium scene, mounted in the wall to look like the view window of an ordinary aquarium tank, except that here you can go behind to discover that nothing is really there. Contrary to what many Disneyland visitors believe, this is the first use of holograms in the complex—all the other amazing effects, including the 'Haunted House', are created with old-fashioned optical illusions!

Holography in Education

The teaching of holography itself in art schools and high schools is now becoming more common, and it is a well-established topic in many science courses at all levels. One of the first independent institutions offering specialist instruction was the San Francisco School of Holography, originally headed by Lloyd Cross, a one-time laser scientist. Two New York artists, Sam Moree and Dan Schweitzer, give occasional 'master' classes in practical holography at their New York Holographic Laboratories. The Massachusetts Institute of Technology now includes holography in its Master's in Visual Science program, under the direction of Stephen Benton. An experimental program teaching holography to practicing English artists operated for two years from 1980 at Goldsmith's College, London, under the direction of artist Susan Gamble and the author.

6 Holography as a Tool

In some ways the field of technical holography has analogies with that of technical and scientific photography. Photography is used by artists, advertisers and newspapers for pictorial display, yet it has certain properties as a technique which enable it to be exploited usefully as a scientific tool; the two fields are not entirely divorced and sometimes a scientific application, such as infrared photography, will be found useful by creative photographers.

The technical uses of holography arise from particular properties of the technique. Just as in photography, a new technical application of holography may find its way into display holography to serve some creative purpose. Many areas of research, such as the improvement of the photographic emulsions used in holography, have had common application in both display and technical holography.

Throughout the 1960s the research groups of many large companies were engaged in analysing different aspects of holography. For many, the motivation for this work was to develop holography as a new kind of technology to improve existing techniques such as electronics and photography. Often, holography was enthusiastically proposed as a solution to some problem for which it was completely unsuitable; inevitably disappointment followed. In fact, among some scientists holography became a bit of a joke—when stumped by a difficult problem the remark would be made: 'Well, we could always try holography'.

One of the most promising proposals is for a device called a holographic computer memory. There are a number of different devices, or systems, capable of storing information in the form required by a computer (for example, magnetic tapes and discs); when creating a new computer the designers will select that memory system which seems to offer the greatest number of advantages in terms of efficiency and usefulness. To date, there have been a number of practical problems, such as the lack of reliable lasers, which has meant that holographic memory systems are inferior to those currently in use. Other technological applications of holography have met with similar problems.

The 1960s were really the heyday of holographic research, when it seemed that as a technique it would receive wide-ranging application to the solution of a diverse range of problems in science and technology. During this period many research groups were involved, particularly in the USA, and hundreds of scientific papers were published on the subject.

Technical applications accounted for the largest part of holographic research in the 1960s and, with their 'failure', many research groups pulled out of holography and turned to other things. Although many groups are still occupied in technical holographic research and development, and work has steadily progressed on display holography, it will be some time before holography realises the wide-ranging application it once promised. Improvements in other technologies such as laser design are necessary before such things as holographic memories become competitive and marketable on a large scale.

Holographic Microscopy

Denis Gabor first conceived of holography while working at the UK Thomson-Houston laboratories, Rugby, on the problem of bringing the magnifying quality of the electron microscope up to its theoretical limit. He published the results of his work in a short letter to the magazine *Nature* in May 1948 under the title 'A New Microscopic Principle'. His invention was not entirely 'out of the blue' and was in a sense, a natural extension of the contemporary research being done on electron-X-ray-microscopy techniques. As well as outlining the basic theoretical properties of holography, Gabor gave a convincing demonstration using visible light from the best coherent sources then available.

The electron microscope is similar in principle to ordinary microscopes which use light. This is because electrons, which are tiny negatively charged particles, also behave as waves and can be focussed by special lenses to form magnified visible images of microscopic objects. Electron waves may interfere with each other and produce the same effect on hitting a photographic plate as light waves, so that regions of a high intensity of electron waves become darkened. Gabor's idea was to use a coherent source of electrons to enable him to produce a hologram of a microscopic object; if this hologram were illuminated for reconstruction with coherent light waves the image theoretically obtained from it would be a vastly magnified 3D view of the microscopic object; this magnification occurs because the hologram is made with waves (electron waves) of a much shorter wavelength than the light waves used to reconstruct it. Gabor hoped that the process would make visible finer details of the object than had previously been achieved with electron microscopes. Unfortunately, there are problems in trying to produce a coherent source of electrons and, in fact, the theoretical quality of such a microscope has never been attained.

However, holography with light waves has found application in conventional microscopy. Light microscopes, such as those used by doctors studying biological specimens, are only ever in focus for one thin horizontal layer through the specimen. Regions of the specimen above and below this layer are out of focus. This can be a problem when trying to observe tiny objects, such as a single bacterium, which are moving in a vertical direction through the specimen, for then one must continually refocus the microscope to follow the movement of the object and maintain it in sharp focus.

Holographic microscopes solve this problem by recording a hologram of the specimen 'frozen' at some instant. We may then examine under a microscope, at leisure, the holographic image of the specimen without having to worry that the object may suddenly move.

A similar application is in studying tiny particles floating in the atmosphere, such as dust and pollutants, which are continually in motion. Looking at a holographic image of a sample of air permits leisurely observation of the size and distribution of the particles at any one instant.

An interesting point about both these applications is that they are examples of a situation in which the holographic image is more useful than the actual object.

Holographic Interferometry

Interferometry is a laboratory technique by which extremely small movements of objects are detected and measured by exploiting the interference properties of light. Movements of roughly the wavelength of light (one thousandth of a milli-metre) can be measured using this technique. Conventional interferometry has a number of limitations which restrict its use to rather specialised applications. An accidental discovery by the first holographers at the University of Michigan indicated that there was a way in which holography and interferometry could be combined in a powerful new technique. Once, when an object being hologrammed accidently moved during exposure, its reconstructed image had a zebra-like pattern of dark lines superimposed upon its surface. The holographers soon realised that the pattern was related to, and in a sense described, the movement which had occurred. They developed the 'accident' into a new technique, called holographic interferometry. Using this technique it is possible to measure the tiny movement produced when an object is slightly distorted in shape, or to analyse the way a surface, such as the back of a guitar, vibrates (Plate 22).

As stated in Chapter 4, it is possible to record two holo-grams on the same photographic plate. What was not stressed at the time was the possibility of interference between the two reconstructed wavefronts. The effect of this is not noticeable if the two wavefronts are quite dissimilar, such as the two wavefronts produced by an object hologrammed in two quite different positions; interference will, however, be noticeable if the same object is deformed very slightly in some way between the two exposures. Typically we would use a deformation so slight as to be imperceptible to the eye; for instance, if the object were held in a vice we could tighten the grip of the vice by a small amount. The result of this interference between two almost identical images of an object is a reconstructed image of the object with a zebra pattern of dark lines across its surface. This pattern describes the way in which the object has moved on being deformed, and is made up of the 'stress lines' of the motion.

19 Double-exposure hologram of a car tyre. Almost microscopic movements of the tyre are made visible as dark lines across its surface—these are contours of equal movement. Mirrors placed at the sides of the tyre enable us to see movements in the walls of the tyre. Two faults (indicated by the arrows) are made visible by the technique.

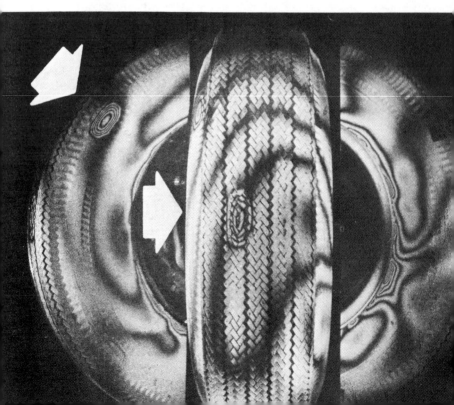

To understand how this pattern arises we must consider the result of the interference between the light waves from the two reconstructed images of each point on the object. If a point moves by half a wavelength or a whole number of half wavelengths in between the two exposures, the two reconstructed waves corresponding to that point will be in anti-phase and cancel each other out on reconstruction: the point will appear dark. Conversely, if it moves by a whole number of wavelengths, the light waves of the two images of it will be in phase and the point will appear bright when the hologram is viewed. The lines of the zebra-like pattern across the surface of the final image thus join points which have moved by the same amount between the two exposures. These lines are 'contours' of equal deformation. In general there will be a concentration of the lines of the zebra pattern around a point where there is some fault or weakness.

The technique is known as **double-exposure holographic interferometry** (the resulting hologram is often referred to as an 'interferogram'), and it can be used for any object which can be hologrammed. It has been found very useful for non-destructive testing of components; for example, car tyres (plate 19). The two exposures are made when the tyre is pumped first to low pressure, then to a higher pressure. Faulty tyres can therefore be discovered without the need to 'test to destruction'.

The technique is particularly useful when used with a pulsed laser, as one can study objects which are moving slowly or creeping; e.g., the growth of a plant over a period of less than a minute. We can see the shock wave produced by a bullet by making a hologram of a blank wall firstly by itself and, then, with a bullet flying across in front of it; the shock wave represents a pressure wave in the air and is made visible by this technique (plate 10).

Another technique, known as **time-averaged holographic interferometry**, is used for studying how surfaces vibrate. At certain frequencies an object such as the head of drum will resonate. When this occurs, the surface is moving in some well defined way; it will be divided into regions which move up and down or in and out, and along the borders of these regions it will be practically stationary. The surface as it moves will spend most of its time at either extreme of its movement, as it slows down to a stop and moves off in the opposite direction. A hologram made of the object *while it is actually vibrating* gives what is effectively a double exposure of the vibrating surface, seen at either extreme of its oscillation. The surface will tend to bulge out or in most at the centre of the

20 Double-exposure hologram of spacecraft microwave antenna. The dark lines are produced as a result of finger pressure applied by one of the men standing to the left of the antenna. The small deformation of the antenna is made visible using this sensitive technique. The jagged nature of the lines above the men is a result of convection currents due to their body heat.

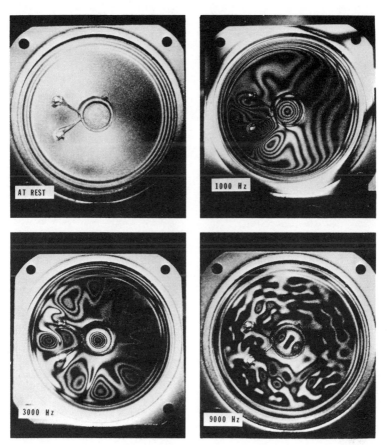

21 Time-averaged holograms of a speaker cone. During an exposure of the holograms the speaker was actually vibrating. The pattern superimposed across its surface describes the way in which the speaker was vibrating at various frequencies and enables designers to improve the quality of sound produced by the speaker.

regions which are moving, though the amount of movement is usually so small as to be imperceptible to the eye. Again, the final hologram image has a pattern of dark and bright lines across it, but now the lines correspond to *contours of equal amplitude of vibration,* and show how much each point on the surface of an object moves when vibrating. This tech-

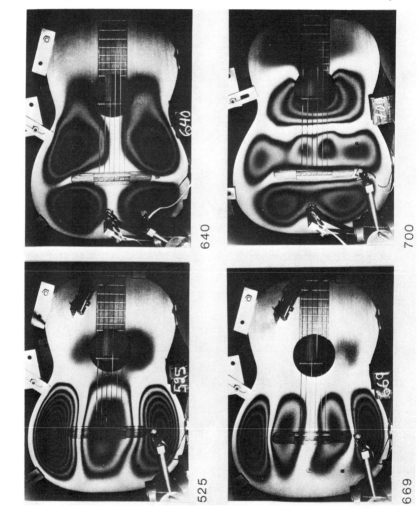

22 Time-averaged holograms of a guitar. These holograms, exposed while the guitar was vibrating at different frequencies, enable the makers to check and improve the quality of sound produced by the guitar.

nique has been used to show the way in which a loudspeaker vibrates (plate 21), and has been used to aid in the design of musical instruments, such as the guitar (plate 22).

To date, holographic interferometry and non-destructive

testing have been the biggest technical application of holography. 'Holocameras' have been produced, using pulsed lasers, to allow the technique to be used outside the laboratory, on the factory floor, where components can be tested as they come off a production line.

Computer Memories

Many large organisations need to store vast quantities of information. Were information to be stored on paper the sheer volume of it would be enough to create quite a problem when it came to retrieving the information in order to use it. There have been a number of technological devices introduced to reduce the physical volume that stored information occupies. At the same time, computers have been increasingly used to process this information; e.g., for producing telephone and electricity bills.

Information can be stored either in a form comprehensible to humans, such as the printed word, or encoded in the binary language of computers as a series of 0s and 1s. In the latter case it can then be automatically processed by a computer. Computers generally use a sophisticated magnetic tape-recorder to store information. To find out a piece of information the computer orders the tape-recorder to wind onto and play a particular section of the tape.

Microfilm is a photographic technique of storing large quantities of information in a small space, and it exploits the capacity of photographic film to record fine detail. To record written information, a photograph is made in which a page of the information to be stored is reduced in size by a lens system so that it occupies only a tiny area on the 'negative' produced. Using this technique it is possible to store 87,500 pages (fifty Bibles) of information on a square inch of film (Collier, 1971). In fact, such extreme reduction in size is not easy to achieve, and even then the smallest spot of dust that finds its way onto the film can obscure some vital piece of information. Using a lesser reduction than this, the system is

117

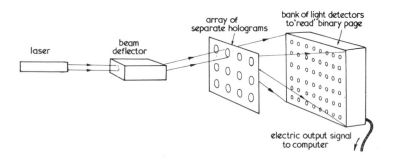

22 Flying spot computer memory. The computer itself chooses which of the array of holograms it will read by directing the laser to the hologram by means of an automatic beam deflector. The binary page stored in that hologram is then read by means of a bank of light detectors.

a very efficient means of storing printed information and is used by many organisations for storing their 'paperwork'.

For use by computers, the information is stored as binary 'pages'. These are square arrays or grids of points at which there is either a spot of light, indicating a 1, or no spot of light, indicating a 0. Such pages can be stored on microfilm and, using an appropriate light detector, the computer can 'read' them. Theoretically, information may be stored more compactly and retrieved more quickly than with a magnetic tape. However, a number of technical problems arise which make a microfilm storage device inferior. A spot of dust on the film could mean that the computer reads a 1 for a 0; unlike a human reader of words, the computer could not easily identify this as a fault and correct it. Another problem is that microfilm forms a permanent record which is not easily changed or updated. To alter one detail on a page requires recording a complete new corrected page on microfilm.

Holography can provide a solution to the former problem by making holographic recordings of binary pages. Since the image of a spot of light is not localised at one point on the film or plate, the difficulty of dust particles is not so severe as with microfilm.

Two types of holographic memories have been proposed. With a volume hologram (see Chapter 4) we may record up to a hundred binary pages as separate holograms on the same photographic plate. Each of them may be read in turn by directing the reference beam at the plate at the correct angle. Plane transmission holograms may also be used in holographic memories. In order to maximise the information stored on the photographic plate, it is divided into a grid of maybe 100 x 100 tiny and separate holograms, each of which stores the image of a different binary page. This is known as a 'flying-spot' holographic memory (fig. 22).

Holographic memories share with microfilm the problem that they are a permanent record. To change part of the memory, to update it with new information, means making a whole new hologram, and this limits the usefulness of the system. It is now possible, however, to use recording media other than conventional photographic plates, parts of which may be erased and 'recorded over' for use in flying-spot memories.

Another problem has been that existing lasers do not have the reliability necessary for use in computers. 'Reliability', in this context, refers to the extremely high standards imposed on devices used in computers, where competition between different devices is very great. (In fact, lasers are manufactured which fulfil the everyday use of the word 'reliable', having estimated lifetimes of up to 10,000 hours. This is comparable with the lifetime of an average lightbulb.)

A French bank has used a volume hologram memory system to store its clients' details. The 30cm-square plate is divided into many tiny volume holograms each 2mm square and each containing sixty holograms. The whole memory can store 60,000 pages of written information.

Incidentally, the holographic process has been suggested as a model to help explain the human memory and how we are able to recognise objects. The argument is an intuitive one, based on analogies between the action of the nervous system, the properties of the human memory, and the making and viewing of holograms of objects using light. With such a model it is hoped to extend neurophysiology beyond the analysis of individual nerve impulses to looking at whole patterns of related nerve impulses. We shall describe this in more detail after the next section.

Pattern Recognition

Holography is a branch of optics, a study traditionally concerned with the design of optical devices using lenses; e.g., cameras, microscopes and spectacles. It is hardly surprising, then, that researchers had the idea of making holograms incorporating the use of lenses. Such holograms, known as **Fourier transform holograms,** have special properties which make them useful for certain technical applications.

In a camera, a convex lens focusses the light wavefront from a distant object to form a flat image on photographic film (see Chapter 1). We are instantly able to recognise it as a pictorial representation of the object. In mathematical terms, the action of the lens **transforms** the object's wavefront into a new wavefront which produces the flat image on the film. The type of transformation which a wavefront undergoes on passing through a lens depends on such factors as the shape of the lens and just where we capture the wavefront on film.

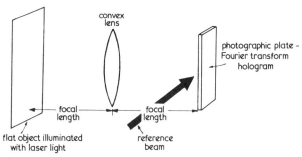

23 Recording a Fourier transform hologram. These special holograms are recorded by incorporating a lens into the system as shown.

In Fourier transform holography, we capture an object's wavefront holographically, after its has undergone a particular transformation known as a Fourier Transform. To do this we place a photographic plate at the focal plane of the lens. We place a flat object, such as a stamp or a transparency, the same distance in front of the lens as the photographic plate is behind it. The object's wavefront, when it reaches the plate, has been Fourier transformed by the lens. We could not recognise the pattern there: it would be quite unlike the original object. If the object is illuminated only by coherent light, and if we provide a reference beam at an angle to the plate, we may record the Fourier transform as a hologram (fig. 23).

A Fourier transform done twice brings us back to the original object wavefront. This provides us with a way to 'decode' the Fourier transform hologram. The processed hologram is illuminated by the reference beam and placed at the focal distance of the convex lens we used to make the holo-

24 A simple convolution.

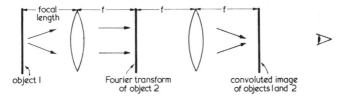

25 Convolution of two objects using Fourier transform holography. The Fourier transform of object 2 acts like a kind of filter to the Fourier transform of object 1. What gets through produces the convolution of the two images. If object 1 were a page of print and object 2 a word which occurred on that page, then the convoluted image would be messy but show a peak of brightness wherever that word occurred on the page. It is hoped that such a technique could be used to produce a machine which can 'read' pages of ordinary printed words.

121

gram. Looking through the lens at the hologram we then see an image of the object in front of the lens. It lies the same distance in front of the lens as the hologram lies behind it.

Fourier holograms may be made which require relatively unsophisticated photographic plates or film to record them. Another advantage is that they enable us to combine two images in an operation known as **convolution**.

Convolution is a rather abstract operation with a precise mathematical description. It has similarities to the operations of addition and multiplication. An example will serve to illustrate the pertinent features. If we convolute three dots with a triangle we obtain three triangles, one at each of the positions of the dots (fig. 24). If either of the images being convoluted is more complicated than a pattern of dots then the effect would be unrecognisable. A triangle convoluted with a straight line, which we can think of as many dots joined together, would result in a triangle 'smeared' along the length of the line.

To convolute two images of flat 'objects' we use the arrangement of fig. 25. The Fourier transform of the second image is recorded on film and acts as a kind of filter to the Fourier transform of the first image. What gets through is transformed by the second lens into the convolution of the two images.

A straightforward application of this technique is in producing several identical copies of a flat object. If object 1 were a stamp and object 2 were three dots, we would see an image of three identical stamps. This technique has been used to produce multiple copies of masks used in the production of tiny printed circuits.

The problem of **pattern recognition** by machine is a field in which Fourier transform holography is potentially useful. The object is to produce a machine which can 'look' at a large, possibly abstract, picture or page and recognise in it the presence of some small pattern or image. A specific and potentially useful application of this is in producing a machine which could recognise a particular word or a particular letter that may occur at some position on a page of printed words. We would then have a device which could 'read' printed information. Of course, a machine such as this would not be able to 'understand' the text unless it were written in the machine's own special language. It could, however, store the text until such a time as someone wanted to read it. It could also be used to transmit the text automatically.

Non-holographic solutions to the problem come up against a number of difficulties. It is very hard to define analytically just what constitutes the shape of a particular letter or the form of a particular pattern. Analytic solutions to the problem using computers get involved in long, time-consuming calculations in which a great deal of information has to be processed.

A convolution technique provides an elegant and efficient solution to the problem. If object 1 is a page of print and object 2 is a letter of the alphabet, then the result is a complicated and unrecognisable convoluted image. However, whereever the letter occurs on the page, there will be a corresponding bright spot of light in the convoluted image, called a 'correlation peak'. By using a bank of light detectors the machine can register and store away the positions on the page where the letter occurs. By going through all the letters of the alphabet in turn the machine will eventually have stored away the positions of all of them on the page. It has 'read' the page, and can store away the whole of it in its memory.

Although attempts have been made to produce optical readers in this way there are a number of very difficult practical problems to be solved: all systems have difficulty in discriminating between the letters 'O' and 'Q'.

The beauty of the technique is that, in effect, it does the same job as the long calculations of the analytical technique. The processing of all the relevant information is done automatically in the convolution technique by the action of coherent light and its ability to form interference patterns.

123

In this respect we can consider the system as an 'optical computer'. In fact, a similar arrangement has been used to process the information received by a radar scanner. Using a conventional computer to do the job would mean waiting for the answers to be calculated analytically, which could take several minutes or even hours.

In terms of technical achievement 'optical computing' or 'optical data processing' is the field in which holography probably shows the greatest promise for the future. Properly speaking, this field is not the domain of 'holography' alone; it calls into play many other branches of optics and physics.

For certain applications the electric currents flowing through individual electronic components in a conventional computer could be replaced by coherent light wavefronts interacting through a variety of optical devices. The great advantages of such systems would be their speed and ability to handle large quantities of information simultaneously; it is possible to do things with optical processing which cannot be done with conventional computers.

The equipment required for optical processing systems presents a number of problems impeding their further commercial development. The optical components of the system, such as lenses, must be very precisely aligned and positioned; the slightest movement could put the whole system out of order. Other problems are with the photographic emulsions used for the holograms and the lack of reliable lasers. As a result, optical processing has as yet been used only in a few specialised applications. More widespread use awaits improvements in other techniques.

Bioholography

We are now in a position to describe some of the more plausible suggestions that certain processes in the brain are holographic in nature. The starting point for these ideas is the known fact that the registration in the brain of a specific piece of information is not confined to one precise point (or neuron) in the brain, but is spread over a region. This is true also in a hologram, where the visual 'information' is spread over the whole of the plate. In cases where people have had part of their brain destroyed they can still remember a piece of information though usually with reduced efficiency; if we cut a hologram into two pieces, each is capable of reproducing an image, though there are deficiencies in each of the reconstructions.

In brief, our ability to recognise certain objects placed in our visual field can be explained using the principles of Fourier transform holography and the operation of convolution. The idea is that the electrical signals received by the brain, every time we see a familiar object, eventually become stored as a Fourier transform **neurohologram**, stored across the same region of the brain as many others; this would subsequently act as a filter to be convoluted with any visual signals we happened to be receiving at a later time. If at any time the object is placed within our visual field, the 'convolution' has a correlation peak of some sort; the object is recognised; and a message is sent to our consciousness telling of this recognition.

The theory is rather controversial and has not gained general acceptance. The arguments are complex and I am not qualified to discuss the issues involved although I think it would be fair to say that there are many workers in the field who feel that it tells a part, if not the whole, truth about these workings of the brain (for more ideas see Greguss, 1973).

Acoustic Holography

As we mentioned in Chapter 1, sound, like light, is a wave phenomenon. In the case of sound, the waves described pressure variations, to which our ears are sensitive, in the air. Sound waves exhibit many of the same wave properties as light, including interference, diffraction and coherence. A pure tone is a perfectly coherent sound wave. Using coherent sound it is possible to produce an interference pattern between an object's reflected sound wavefront and a coherent reference sound wave; this will be a pattern of pressure in the air. However, if by some means we can make it visible and record it, then we can use it as a hologram to be illuminated with light and so obtain a visible image of the object. Such a hologram is known as an **acoustic hologram**.*

Sound waves travel through water just as they do through air, and for various reasons it is easier to form acoustic holograms in water than in air. A simple arrangement for forming an acoustic hologram is shown in fig. 26. Coherent sound of high frequency is produced by two **sound transducers**; metal discs which vibrate at a frequency of about five million cycles per second. In optical holography both the object's illuminating beam and the reference wavefront have to be derived from a single laser; the equivalent process is not necessary in acoustic holography since sound transducers can be made which are perfectly coherent, and two separate transducers can be used to provide the two sets of waves.

One transducer is pointed at the object to be holographed and the other at the surface. The interference pattern between the object's waves and the reference wave is formed on the surface of the tank as a stationary pattern of ripples. These may act as a phase reflection hologram to a laser beam reflected from the surface. Because of the difference in wavelength between the sound waves and light waves the image of the object appears to be a long way beneath the surface and must be viewed through a telescope. An advantage of this system is that, as the object moves, the hologram changes accordingly and we may follow the movements of the object through the telescope. There are a number of improvements of this simple technique and it has been extended to give images of objects in air.

Acoustic holography is potentially useful for giving under-

---

*Properly, acoustic holography is a subject in its own right; however, we have chosen to include it here because its applications are largely technical.

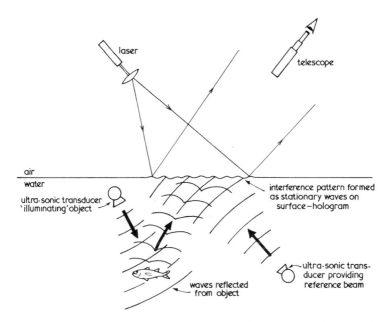

26 Acoustic holography underwater. Coherent sound waves in water, of very high frequency, are capable of producing an interference pattern of ripples on the surface of the water which, when illuminated by laser light, acts as a phase reflection hologram to produce a visible image, which can be viewed through a telescope, of an underwater object.

water sight to submarines. A system has been developed to do this which has a range of several hundred metres and can be used for navigating through difficult waters to search for and rescue other submarines and to map the sea bed.

Acoustic holography may be used also in medicine to complement X-rays. For some time now ultrasonic scanning techniques have been used in medicine to produce 2D cross-sectional pictures of the human body. However, using acoustic holography it should eventually be possible to produce high-quality 3D images of the internal structure of the body, showing organs and blood vessels. This would be a very useful diagnostic tool.

It has been suggested (Greguss, 1973) that dolphins may

use acoustic holography as a perceptual sense. It is known that they produce 'bleeps' of sound of the requisite frequency. They also possess an organ, known as the melon, which may be sensitive to acoustic interference patterns (i.e., holograms) formed within its volume. The bleeps they emit could simultaneously pass through this organ, providing an acoustic reference beam, and be reflected from surrounding objects back to the melon where they would interfere with the reference beam to form a volume acoustic hologram (fig. 27). The dolphin would perceive this pattern directly and so be able to form an awareness of the objects nearby. Just what this sense would be like subjectively is impossible to imagine.

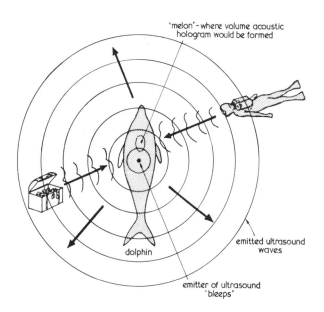

27 Possible holographic sense in dolphins. It has been suggested that dolphins, by emitting a high-frequency 'bleep' of sound, may be able to form, within an organ known as the melon, a volume acoustic hologram of the surroundings. They may thus be able to be simultaneously aware of the shape, form and distance of two objects in opposite directions.

As with sight, it would be possible to determine the position and distinguish the shape and form of any particular object; but, unlike sight and rather like hearing, it would not be directed at a single object: it would be possible for the dolphin to be simultaneously aware of two objects in opposite directions. Tests of this hypothesis have included such experiments as showing that dolphins, while blindfolded, have been able to distinguish between a grid of vertical bars and one of horizontal bars. In fact, it is difficult to prove decisively that such a distinction is directly attributable to the type of sense proposed.

7 The Future of Holography

In his science-fiction novel *The Eighty Minute Hour*, Brian Aldiss portrays a future world in which a young girl is kidnapped by a holographic doppelgänger of her mother. The folk heroes of this age are those individuals who use their imaginations to create for the masses 'holodreams'—fantastic three-dimensional experiences. These ideas indicate some of the more powerful uses to which holography may be put in the distant future. Perhaps one day it will be possible to talk to a three-dimensional real-as-life image of someone who is many thousands of miles away—a partial realisation of the science-fiction writer's dream of 'matter transmission', in which people are dematerialised and 'beamed' to a distant location. Many of the developments in technical holography which we described in Chapter 6 only await advances in other technologies to become commercial propositions.

In the near future we may expect to see the further implementation by advertisers, educationalists and artists of existing techniques in display holography (Chapter 5). Widespread use by advertisers on billboards and at other locations, such as 'point of purchase' displays, may render the hologram a commonplace artefact of everyday life. As mass-production techniques are developed, the cost of holograms will gradually be reduced. Pictorial reflection holograms, viewable in direct white light, may soon replace the china ducks and mass-produced paintings which decorate many living-room walls.

As the quality of lasers improves and with the development of new techniques we can expect to see larger holograms of more extensive, even panoramic, scenes. These could be particularly useful in allowing flexible and realistic scene changes in the theatre, affording heightened realism. The audience could actually be placed in the centre of the scene, surrounded by holographic projections. Holographic auditoriums have been suggested which would combine the projection of large-scale panoramic holograms with a laser light show. Large white light reflection holograms could be used to adorn otherwise drab office blocks and industrial estates; they could be used to provide panoramic views of exotic locations in air terminals and restaurants. Countless other uses have been suggested; for instance, a hologram of a police car planted near a busy road might cause passing motorists to ease their foot off the accelerator.

Another development, which lies somewhat further in the future, would follow the invention of an X-ray 'laser' which emitted coherent X-rays instead of light. X-rays have a much shorter wavelength than light and an X-ray laser would permit us to form a hologram of a microscopic object, perhaps even a molecule, which when illuminated with visible laser light would create a vastly magnified three-dimensional image of the object. Such a device would be a tool of inestimable value to scientists investigating the basic microscopic structure of matter. Unfortunately, the problems of producing an X-ray laser are extremely great, one being the need for a huge power source to supply the laser with energy. Much research is being devoted to the production of such a device, and there is some hope that these efforts may be rewarded in years to come.

There has been considerable speculation, and much research done, into the possibility of creating moving holograms; i.e., holograms which capture and then recreate the action which occurs in a scene. The ultimate in realism would be moving holograms with sound and in colour. Both holographic movies and holographic television have been suggested.

To understand how these might work we must first consider the general principles of ordinary movies and television. These work by presenting on a flat screen of some sort a succession of 'frames' or still-life pictures, one approximately every one thirtieth of a second. At this rate, any movement which occurs

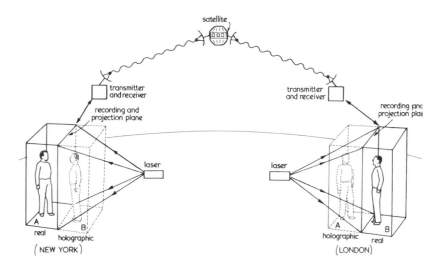

28 Scheme for a holophone link. A holographic television may, in the distant future, provide a partial realisation of the science-fiction writer's dream of teleportation. Such a system, which might be known as a holophone, would allow A in New York, to talk to a life-like 3D image of B, who is in London. A and B would enter identical booths in both towns, one wall of which would be capable of recording their interference pattern for transmission, and simultaneously have projected on it their partner's interference pattern which, when illuminated by laser, would recreate a 3D image of the partner, who would appear to be sitting on the other side of a glass wall.

in the picture is recreated smoothly and evenly, without jerkiness. In movie films this process is accomplished by making a reel of film consisting of a series of colour (or black-and-white) transparencies which are then projected in rapid succession onto the cinema screen. In television the individual still-life pictures are produced by a spot of light which scans repeatedly and at high speeds the lines of the picture tube (625 in the UK, 525 in the US), varying in intensity from point to point according to the brightness of the picture at each point and according to the signals sent from the central television transmitter. The spot moves so quickly that we never see it, and it creates many still-life pictures each second.

By **holographic movie** we mean a series of holograms stored permanently on film and projected (i.e., illuminated by laser) every one thirtieth of a second. By **holographic television** we mean the transmission from central stations of hologram programmes to be received by and viewed on a device in people's homes; or, alternatively, a two-way line 'holophone' to enable individuals or groups to 'meet' over long distances. It has been suggested that by setting up holophone communication links it would be possible to conduct international business and political conferences without the need actually to send the persons involved abroad. Such a link would obviate the need for many of today's business journeys. A holophone line in the home would in the same way reduce the need for many other sorts of travel—e.g., to work.

In a holographic TV system we would need to be able to transmit to each home sufficient information to construct in a device there roughly thirty holograms in each second. As we have already mentioned, holograms contain a lot of information, a fact which can be exploited in a device such as a holographic computer memory. Any TV channel has a certain limit to the rate at which information may be transmitted through it (by one of the laws of communications science). A hologram contains as much as 300,000 times more information than an ordinary television picture. Using a normal TV channel it would take well over *two hours* to transmit sufficient information to produce *one* 10cmx10cm hologram. Obviously this is not fast enough, since we want thirty every second to create movement. If we won't mind sacrificing some of the quality in the image, we can reduce the amount of information contained in the hologram. For instance, we could dispense with vertical parallax in the image, as is done in the rainbow cylinder hologram (Chapter 5). Even then, however, the information content is still too high.

Leaving aside the abstract consideration of information content, there are a number of technical problems which holographic TV systems come up against. Having received the information in the hologram's interference pattern it is necessary to devise some means of impressing it on a light-sensitive screen which may then be illuminated by a laser or other form of coherent radiation contained within the television set to form the holographic image.

The many problems of holographic TV systems mean that it will be a long time, probably well into the next century, before they can be introduced into widespread use. The problems appear so hopeless that at present there are no firms willing to fund the necessary research. Long before the advent of holographic 3D television we can expect to see the introduction of a 3D TV system based on the stereoscopic principle (Chapter 1) probably using a screen which allows the TV to be viewed without special spectacles.

Many ingenious attempts have been made to produce three-dimensional movies. Before holography, all of these used some variant of the stereoscope principle, which creates the illusion of seeing in 3D by sending dissimilar 2D pictures to each eye. There were a number of difficulties, including the need often to wear special spectacles, which prevented such systems gaining widespread popularity. Holography is capable of producing truly 3D images, and it is hoped that one day it may be used to bring 3D movies to the commercial cinema.

One of the earliest holographic movies made use of the fact that volume holograms are capable of storing many separate images on the same photographic plate (Chapter 4). Each of these images can be one 'frame' from a sequence which makes up the holographic movie. To view these in the correct succession either the plate or the illuminating light beam is rotated slowly through an angle. As the beam hits the plate at increasing angles it causes the separate images, or frames, to be illuminated in sequence. To make such a movie, the original reference and object illuminating beams are derived from the same, pulsed, laser and are strobed (or 'flashed') at a rate of thirty times per second while the plate is slowly revolved. There are a number of limitations to this technique which have meant that it has not been further developed. Since we may only record about 100 holograms on each photographic plate, and we need 30 frames per second for a movie, then the total length of the movie will be only about three seconds. This is acceptable if the motion we wish to record is an animated sequence which for demonstration purposes we could view at a slower rate than thirty frames per second. It means however that such a system would not be useful for holographic cinema. Another problem is that, since we are using a photographic plate of relatively small size (say 25cm^2), this limits the size of the window through which we can view the scene being reconstructed and consequently limits the number of people who could simultaneously watch such a movie.

As mentioned in Chapter 5, the rainbow cylinder hologram technique may also be adapted to record motion. However, only sequences less than forty-five seconds long may be recorded, and so, again, the system would not be of use in a holographic cinema.

Another system, developed more recently in the USA and the USSR, while allowing the recording of action scenes of longer duration, is hampered at present to an even greater extent by the size of window problem. In this system the sequence of holograms is recorded as rectangular frames on a long reel of film 7cm wide. On viewing, the film passes by continuously in front of the observer. As each hologram arrives at the position of the imaginary viewing window it is illuminated briefly by a flash of light from a pulsed laser. In each second, thirty holograms are illuminated in sequence as they arrive at the position of the window. By looking through the window the observer can follow the motion of objects in a truly 3D scene. An early attempt at this kind of system was a short film, lasting less than a minute, of some goldfish swimming in a tank (Jacobsen, 1969).

To record the sequence of hologram frames on the film the reference beam (to the film) and the object's illuminating beam

135

are provided by a pulsed laser. The laser is strobed thirty times per second while the film moves continuously past the scene which is the subject of the movie (fig. 29). During the very short duration of the pulses the film hardly moves at all and each hologram frame is recorded sharply without any blurring of the interference pattern. Thus, in every second, thirty separate holograms are recorded in succession on the film.

How long the movie lasts is limited only by the length of the film, but so far such movies have lasted not more than a

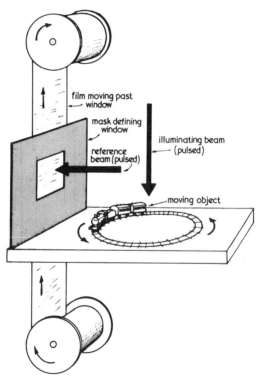

29 Scheme for a holographic 3D movie; an arrangement similar to that shown here has already been used to produce simple holographic 3D movies. On reconstruction the observer looks through the window defined by the moving film and sees behind it the recorded scene, complete with action. This system may be improved in a number of ways.

few minutes. Since the rectangular holograms on the film are only 7cm wide the window through which the scene may be viewed is tiny; it is difficult for more than one person at a time to view the movie. It has been suggested that this difficulty could be circumvented by having a continuous loop of film which passes from one viewing device to the next in a purpose-built cinema, thus giving a staggered viewing at each of the devices. Also, the technique is being extended to use film which is much wider, and at some time in the future huge reels of film may be produced which are at least 1m wide, enabling hologram movies to be produced which have a viewing window larger than the screen of a normal television. This will greatly enhance the 3D effect of the movie and also enable a larger audience to view the movie at one showing. It has been suggested that eventually the size of the film could be increased to such an extent that the window were the same size as a conventional 2D movie screen. Alternatively, a large lens could be placed in front of the window of the film, producing a magnified image that a large audience could see.

Such movies are restricted by the same limitations as occur with still-life display holograms. The recording must be made in a darkened room or studio so that the only illumination of the scene being recorded is coherent laser light. The image suffers from speckle as in display holograms. Also, it will be some time before this type of holographic movie can faithfully reproduce the natural colour of objects. These limitations mean that for some time to come holographic movies will be inferior in most respects to conventional 2D movies. Research is underway to solve these problems, and fully 3D coloured holographic movies may well become a commercial possibility by the end of this century.

It might seem at first sight that holographic movies would have little to gain over real life productions on the stage, where the actors and props, being physically real, are of course fully three-dimensional. However, the versatility of the holographic medium ensures that it will be able to create a far more realistic atmosphere than a conventional stage. It would be possible to shoot scenes on location using real buildings and landscapes as scenery. It will be possible eventually to completely surround the audience by an artificial three-dimensional scene; the audience could thus be placed right in the middle of an exciting scene full of action. Scene changes could be facilitated in a holographic movie with the same ease as in a conventional movie; one could 'cut' from one scene to the next. To heighten the illusion of reality, holographic cinemas could be fitted with sophisticated stereo sound systems.

Thus audiences of future Westerns may be placed, apparently, at some convenient part of the desert from which to watch the plot unfold. Computer-generated holographic movies would open up a whole new medium in creative expression and entertainment enabling totally synthetic visual experiences to be created.

It is a terrifying thought that life-like 3D movies could also be used in the future for psychological conditioning or brainwashing.

8 Making Your Own Holograms

Having read about holography in the previous chapters you may have decided that you would like to attempt to make holograms yourself. For the price of a good-quality conventional camera system it is possible to obtain all the necessary equipment to produce your own holograms, and the technical skill and expertise required for holography is comparable to that of amateur photograph. Making holograms can be simple and fun.

The holographic set-up pictured in plate 7 represents a degree of sophistication (and expense) which is not needed for the production of many of the simpler types of hologram. When contemplating what sort of holographic set-up to use you have a choice—either you buy an 'off the shelf' ready made 'holographic laboratory', with which it is claimed you can make your first hologram within an hour, or, with the investment of a little time and trouble, you can make the components and equipment for your own holographic set-up.

At least two firms in the United States supply ready-built 'kits' to make holograms. Remember that whereas prospective purchasers of photographic equipment can look at the goods with a helpful sales assistant in a local camera store, it will be virtually impossible for you to get to see and handle holography equipment, so study the manufacturer's literature carefully before you make any purchase. Best of all, try and find a nearby introductory practical class with a holographer before you buy equipment. Make sure if you do buy a kit that it will make the holograms you want, and be aware that your interest could quickly outgrow the capabilities of the most basic kits, which may be designed for schools, where the object is simply to allow each student in a class to make a simple hologram.

A number of ready-made holographic kits are produced by Metrologic Instruments Inc. of Bellmawr, New Jersey (see addresses list). The cheapest of these is a basic cylinder-hologram set. Metrologic also produces the somewhat more expensive Advanced Holography Laboratory, which can be used to conduct a number of holographic experiments, including interferometry.

A more advanced—and expensive—range of kits is supplied by NRC Newort Corp. of Fountain Valley, California. These are based on an 'optical breadboard' system which the company makes for professional optics laboratories and is among the best of its type. Newport has also produced a useful manual 'Projects in Holography' to go with its beginner's kits.

The advantages of a do-it-yourself method are cheapness and versatility. Versatility is an important consideration since ultimately the quality of the holograms you produce, and the satisfaction you obtain in so doing, depends on your own ability to adapt and invent, constantly improving the system you use.

Our comments on practical holography will be directed mainly towards those who wish to construct their own holographic equipment, though many of the details will be found of use to anyone opting for one of the ready-built systems. In Section I we shall discuss the equipment needed and suggest a number of different ways of setting it up; In Section II we shall describe a few simple ways of making transmission and white light reflection holograms; and, finally in Section III, we shall suggest ways to improve your set-up and some interesting experiments you might like to try.

For those who have mastered the experiments described here, or who require a more detailed explanation of the technique, other texts on the subject, technical and non-technical, are listed in the bibliography (page 166).

Section I Darkroom & Equipment

As described in Chapter 4, holograms are made by allowing laser light reflected from an object (or objects) to interfere with a simple laser beam, the reference beam, forming an interference pattern which is recorded on special photographic plates or film. Placing the processed hologram in the reference beam causes it to reconstruct, by diffraction, the object's reflected wavefront of light. Basic to any holographic set-up is some means of expanding the narrow beam which emerges from the laser, and some device by which part of the beam can be directed onto the object to illuminate it while the rest travels more or less directly to the plate as the reference beam.

Two major problems occur in making holograms. Firstly the apparatus must be isolated from vibration, which would blur the recording of the interference pattern, resulting in a partial or complete loss of the image. Secondly, because the interference pattern consists of very fine detail, special photographic plates and film, made specifically for holography, must be used.

Darkroom

The darkroom can be any room in your house capable of being blacked out completely. This is necessary both for making the holograms (so that the only light affecting the plates is the coherent light originating from the laser) and for processing the plate. There must be no strong draughts in your darkroom as these could cause some of the equipment to move during an exposure. The floor should be as firm as possible, preferably a concrete floor such as a ground floor, basement or even the garage floor; structural vibrations in your house caused by footsteps or passing traffic could cause enough movement to ruin the holograms. It should also be quiet in your darkroom, with no movement or talking while a hologram is being exposed—sound can cause the apparatus to vibrate. Changes in temperature in the room will cause components to move, by expansion and contraction, and should be avoided if possible. If your darkroom is so cold that you find it uncomfortable to work in it without some form of heating then switch this on at least an hour before you intend to make holograms to allow time for everything to warm up. Also avoid opening the door—letting in a cold draught—just before you expose a hologram.

Laser

The laser will certainly be the most expensive part of your equipment. You will need a helium-neon laser with a power output of one half to one milliwatt. As mentioned in Chapter 3, light from such a laser hitting your skin is quite harmless but *great care must be taken at all times that light coming directly from the laser does not enter your eyes,* since this could cause permanent damage to the retina. In particular, when viewing holograms, although the light in the reconstructed wavefront, which emerges at an angle to the reference beam, is quite harmless, you should be very careful not to look straight down the reference beam at any time.

Lasers which would be suitable for holography are:

Laser	*Power* (milliwatts)
Metrologic M–600	0.5
Metrologic ML–620	0.9
Spectra-Physics 155	0.5

You may want to buy a support for the laser from the manufacturers or simply mount it on a solid support such as a stack of bricks. The Metrologic ML–620 laser has a special mounting to accept the manufacturer's spatial filter, a device which removes annoying interference patterns present in any laser beam. We shall describe in Section III (page 162) how you can make your own spatial filter.

If you have some experience in practical electronics you may be interested in the prospect of building your own laser, thus saving yourself a substantial amount of money. Both kits and plans for circuit designs are available and laser tubes may be purchased ready-made.

The laser should be switched on several minutes before making any holograms to allow it to warm up, after which it will settle down to stable operation.

Optical Components

In most simple holographic set-ups the beam is first expanded in size, then split into two beams which may be directed to their appropriate positions by reflection from mirrors. According to the set-up used, these components are supported by appropriate holders, described in the next subsection.

All that you need for a beam expander is a convex glass lens with a focal length of about 1cm or less and a diameter of about 1cm. Lens quality is not important; suitable lenses are available from Edmund Scientific. On passing through this lens, the narrow beam of the laser is focussed to a point beyond the lens, from which it spreads out as a cone-shaped beam which provides a circular area of illumination if it is allowed to hit a flat surface. The part of the beam which is to be the reference beam must be wide enough by the time it reaches the photographic plate to illuminate the plate uniformly across its surface. If it is too narrow, the circle of illumination may be made larger either by moving the laser further away from the plate or by using a beam-spreading lens with a shorter focal length, which causes the beam to be spread out through a larger angle.

Splitting the beam can be done in several ways. In some set-ups we may use, say, the left-hand half of the expanded beam to light up the object while the right-hand half is used as a reference beam. But often this is not satisfactory, since the cross section of both beams is then a semicircle, and it is difficult to illuminate the plate with uniform intensity with a reference beam of this shape.

A better arrangement is to split the expanded beam by passing it through a glass plate tilted at an angle to the beam. Most of the beam will pass straight through the glass plate but some will be reflected from the surface and off to one side where it can be routed by mirrors to an appropriate

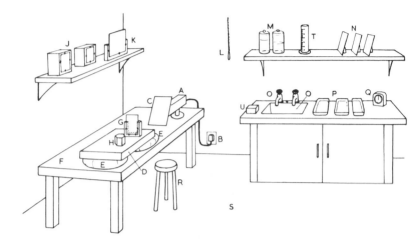

30 A holographic studio. This is a type of set-up which could be used by an amateur to produce holograms. Pictured here is an arrangement using a paving slab vibration-free table (page 146) ready to make a white-light reflection hologram (page 155)— probably the simplest type to make but the most prone to failure by vibration. On the shelf above the table are a few components which could be used to make the other transmission holograms described in the text. Items in the studio are:

A   laser, with beam spreader lens obscured by C;
B   power point for laser;
C   black cardboard shutter leant against the laser;
D   paving slab vibration-free table;
E   partially inflated motorcycle inner tubes;
F   solid, heavy bench with firm legs;
G   photographic plate holder—a U-shaped piece of wood;
H   object being holographed—a hexagonal post;
I   photographic plate;
J   front surface mirrors mounted on blocks (not in use);
K   beam splitter mounted (not in use);
L   convenient pull-on light switch;
M   processing chemicals (developer and fixer);
N   processed plates propped up to dry;
O   sink for washing plates;
P   dishes for processing chemicals;
Q   timer with hands visible in the dark;
R   stool;
S   solid floor, preferably concrete;
T   measuring flask for chemicals; and
U   light-tight container for photographic plates

position. Unfortunately, there is reflection from both the front and back surfaces of the plate and these two reflections travel off in the same direction and may interfere to produce a pattern in the beam. If we use this beam as the reference beam (which is often the case) this interference pattern, which consists of vertical bars, will become superimposed across the hologram which will consequently seem to have a grid of these bars across its surface.

We can make the reflection from the back surface negligible by thinly silvering the front surface of the beam splitter (silvering is the process which turns glass sheets into mirrors; in fact, aluminium is used in preference to silver). This makes the reflection from the front surface much brighter in comparison with that from the back while still allowing a portion of the beam to pass straight through. The best beam splitters are those made from extremely flat 'float glass' which has a thickness of about 6mm. A convenient size is a square sheet with an edge of about 13cm. Float glass is available, in the UK, from Offard Scientific Equipment Ltd, and, in the USA, from Edmund Scientific (see Addresses) and may be obtained 'half-silvered', in sheets of any size.

Expanded laser beams may be routed by reflection from mirrors. Normal mirrors are coated with metal on the back surface. Some reflection also occurs from the front surface of the glass and, as with the beam splitter, this reflection may interfere with the light reflected from the back surface to introduce an interference pattern into the laser beam. For this reason it is best to use 'front-surface' mirrors which have a complete coating of aluminium on the front rather than on the back and so produce only one reflected beam. For best results these too should be made of 6mm float glass. Again, a convenient size is about 13cm^2; such mirrors may be ordered from the same suppliers ·

Great care must be taken when handling front-surface mirrors and beam splitters since the thin layer of silvering is easily damaged. You should never touch the silver surface since this will leave a fingerprint on it. After some time dust may collect on the surface: this is not too important but, if you do wish to clean the mirrors, this may be done by washing them in lukewarm water with a little detergent; vigorous rubbing of the front surface should be avoided since it will remove the metal coating.

Vibration-free table

The main requirements of a vibration-free table for holographic work are that it should be heavy and isolated from movement and vibration. The optical components described above will be supported by holders designed for use with a particular type of table. In designing holders you should always aim for solidity and weight in order to keep the components absolutely still.

Perhaps the simplest type of vibration-free table consists of a concrete block with a top about 60x40cm or larger (e.g., a paving slab) on one (or better two) partially inflated motorcycle inner tubes. The tubes effectively insulate the slab from all but the very largest vibrations or movements in the floor. This table may be placed either on a large stout bench or directly on the floor.

If you use such a table the mirrors may be supported by attaching them to solid wooden blocks with clips gripping their edges. Suitable clips may be simply made by passing a wood screw through a wide washer and screwing this assembly into the wooden block next to the mirror. Alternatively, they may be glued to the blocks using an epoxy adhesive, although this obviously makes the assembly a permanent fixture.

To mount the beam splitter you must use a U-shaped piece of wood so that some of the light may pass straight through; this may be made from a thick piece of plywood cut with a fretsaw.

The plate holder may be made from a wooden block in such a manner that the plate rests along its bottom edge on two metal pins or nails while being held at the top by spring clips. Spring clips may be fashioned from stout pieces of suitable wire (available from modelling shops) covered with insulating tape and screwed to the holder. Remember that you will have to place the plate in the holder in absolute darkness, so design the holder accordingly. If you want to make flat white light reflection holograms you need to make the holder from a U-shaped piece of wood so that the reference beam may be introduced from behind.

The beam-spreading lens may either be mounted directly onto the laser or set in plasticine inside a short section of plastic tube (e.g., plastic drain piping) of appropriate diameter, the assembly then being firmly attached to a wooden block.

When arranging the components in their correct positions for a particular set-up, the blocks holding them may simply be rested on the concrete slab. Their positions having been determined—before exposing the hologram—they may be attached to the table more firmly using molten wax or a suitable temporary adhesive.

If you have access to a metalworking workshop, components such as mirror holders, plate holders, etc., could be made from mild steel. They will then be more rigid than wooden ones and heavier, which helps to increase their stability.

To prevent unwanted reflections from the components in your apparatus, they may be painted with matt black paint.

Another vibration-free table and support system you may wish to consider is the 'sandbox'. Take a stout wooden box at least 15cm deep, line it with a large plastic sheet, place it on one (or two) partially inflated inner tubes and fill it with sand (available from a builder's merchant). All the optical components are then mounted on short, solid wooden posts with pointed ends that can be driven into the sand. Held quite firmly by the sand, which also provides the weight of the table, they may be moved about easily. Such a system is extremely versatile and particularly useful to the amateur since it is possible to make a quite large table which would be impossible to lift were it all in one piece; for instance, if you built walls around the edges of an old door laid flat you could make a sandbox table of a very useful size (you would need four inner tubes). The components may be mounted on blocks similar to those used for the concrete slab table but with the addition of wooden or metal spikes.

With a little bit of ingenuity, plastic drainage pipes can be adapted to act simultaneously as holders and posts. If you cut a slit down the length of the pipe the same width as the thickness of the mirrors and beam splitters, those components can be fitted into the slit and held firmly. The pipe may then be driven into the sand. If you fear that by this means part of the mirror or beam splitter will become obstructed then choose a bigger mirror and don't count on using the section which is obstructed by the pipe. The components should be mounted as low down as possible, even touching the sand, to help minimise any vibration. Plate holders are probably simpler still to make out of wood, but perhaps you can think of your own method which might be better. If using a sandbox, you may also require some sort of platform on a post on which to place any objects you wish to hologram. Again I suggest that you use wood to construct such a device. Alternatively you can rest the objects on the sand surface of the table. (Allow 20 min for the sand to settle.)

Unless your vibration-free table is very large, you may find that there is not enough room on it for the laser as well as the other equipment. In fact, vibration in the laser is not as important as vibration of the other equipment so it would be all right to mount the laser either on another part of the bench, if your table is on a bench, or on another part of the floor if you have placed your table there; however, it is always preferable to place the laser on the vibration-free table if you do have room. If you do place the laser on the vibration-free table you may find that vibrations are transmitted to the table *via* the laser's electric power cable, although this is unlikely.

(If it should occur, try suspending the cable from the ceiling.)

If you find that, despite your efforts, the table is not steady enough to produce holograms then you should increase its weight and improve the isolation from vibration; if your equipment is mounted on a table or bench, the latter can be done by placing the legs of the table in buckets of sand.

If you intend to use flexible photographic film instead of glass plates (see below), this may be firmly held by sandwiching it between two pieces of float glass of the appropriate size.

Photographic plates and film; developing; making an exposure

To record your holograms you will need to use plates and film in the Holotest range manufactured by Agfa-Gevaert. These are coated with a black-and-white photographic emulsion specifically designed for use in holography and capable of recording the fine detail of the holograms' interference pattern. The '75' range is intended for use with the helium-neon lasers I have recommended, and is available in two sorts, 10E75, for transmission holograms, and 8E75, for reflection holograms. Both of these are available on glass plates sized 9x12cm (box dozen). Up-to-date prices are available from the suppliers (see addresses). The 10E75 type is also available for less money on flexible sheets of plastic film 4x5in in size. It is possible, in the future, that some changes may occur in range: for latest details write to Agfa-Gevaert (see addresses, page 165). All of these plates and film are available on order either through your local Agfa photographic dealer or from Integraf. To begin with you will find it easier to use the smaller-sized plates.

Both the plates and the film are processed in exactly the same way as ordinary black-and-white plates or film, no special chemicals being needed; any universal developer (such as Ilford PQ or Kodak equivalent) and standard fixer (such as Ilford Hypam or Kodak equivalent), purchasable at photographic dealers catering for amateur photographers, may be used to process them.

We shall describe only briefly the processing technique; for further details regarding the mixing of chemicals, etc., you may refer to any simple text on amateur photography. You will need three shallow dishes, one for developer, one for a stop bath or water, and one for the fixer. Place in each enough liquid to cover the plates or film, at the correct temperature of 20°C. A timer visible in the dark is required to process the plates (no safelight can be used) or a helpful friend who will stand outside and knock on the door at the appropriate times. Another timing method is to record a tape for a tape recorder, which includes cues such as 'take the plate out of the developer' at the appropriate moments.

After exposing a photographic plate (hologram) you may store it away in a light-tight container (such as a thick black plastic bag with no holes) or wrap it carefully in black paper, then switch on the lights and prepare the chemicals in their dishes. Switch off the lights, and place the hologram, emulsion side up, in the developer. You can tell the emulsion side because it feels less smooth than the opposite side (glass) and because it will stick slightly to your lower lip. Rock the developing tray for the appropriate time (4min with Ilford PQ) and then remove the plate, wash it thoroughly in the stop bath and place it in the fixer (4min with Ilford Hypam). After fixing you may switch on the lights and examine the hologram which, if properly exposed, should, from a distance, appear uniformly grey. It must then be washed in running water for half an hour. Just before removing it from the wash gently smear a few drops of detergent across its surface with your fingers, wash this off and then hang or prop it up vertically, on one corner, in a dust-free place to dry. After this the hologram should be ready for viewing.

To economise on plates and film, especially in your early experiments, you may wish to cut them in half. With film this is quite easy using scissors; this must be done in absolute darkness, so practice first with a piece of paper. To cut plates in half you need a diamond-bit glass cutter and ruler (available from a DIY shop). This is quite difficult, so practice in the dark with similar-sized glass sheets first. You may have to adapt your film holder or make another to accept half-sized glass plates.

You will need also some sort of shutter for the laser. This need not be very sophisticated—a piece of black cardboard leaning against, and covering, the output aperture of the laser will do. With the shutter in place, the plate or film may be put in position. Although some light may leak from behind this simple shutter it should not be enough to fog the hologram. To make an exposure, wait a minute or two for the apparatus to settle down and then carefully lift the cardboard shutter away from the laser for the desired time, making sure you do not knock any of the apparatus in doing so. The exposure may be timed either using a watch with luminous second hand or by counting the seconds to yourself, since exposure time does not usually have to be too exact.

The exposure time will depend on the power of your laser and on the nature of the particular set-up you use. It will normally be somewhere between one and ten seconds. With any particular set-up, once you have determined the correct exposure time for a hologram of one particular object, this time can be used for any other hologram of other objects which you wish to make with the same set-up.

The correct exposure time can be determined only by trial and error, but fortunately, as we have mentioned, it is not too crucial and you should be able to obtain some sort of result with exposures of widely differing durations. As an initial experiment with any set-up try the following exposure times with different plates: 8 sec; 2 sec; as short as possible. One of these should give you some sort of result.

The photographic plate or film, if properly exposed, should be a similar shade of grey to a normal black-and-white camera negative. If it is darker, the exposure was too long; if lighter, too short. The best thing to do is to compare the quality of images produced by the different holograms and then decide which is the best. If you are familiar with the test-strip technique used in determining exposure times for photographic prints you might like to adapt this for the hologram by masking off different sections of the plate with a piece of cardboard, but be careful not to move the photographic plate in doing so. In the unlikely event of the set-up requiring a shorter exposure time than you can make with the cardboard shutter, you may reduce the intensity of your laser beam by placing in its path a neutral grey filter (available through a photographic dealer). The optimum exposure time will thus be increased.

Section II Making Holograms

Below are a number of different arrangements you can use for making both transmission and reflection holograms: this is not to be considered as an exhaustive list of different techniques, but rather as a few suggestions which will give you experience in making holograms and provide you with ideas which you may wish to incorporate in set-ups of your own invention. The basic elements of any set-up are to provide an object wavefront and a reference wavefront which interfere on the photographic plate or film. With a transmission hologram both of these hit the plate from the same side at an angle to each other; with a reflection hologram they hit the plate from nearly opposite directions—i.e., one from the front and one from the back. There are many ways in which this can be achieved.

Simple transmission hologram with no beam splitter

The essential features of this set-up are shown in fig. 31. The components must be arranged as shown, but to do this satisfactorily a number of details must be attended to. The laser and beam spreader must be far enough away from the

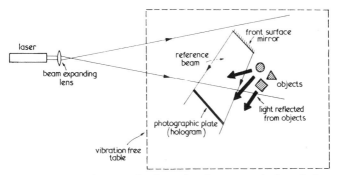

31 Arrangement for producing simple transmission hologram (page 152).

rest of the apparatus for the beam to be large enough to illuminate the whole of the object and, by reflection from the mirror, the whole of the photographic plate. To check this, place a piece of white card, the same size as the photographic plate, in the plate holder.

Another important consideration, with this set-up as with the others we have yet to mention, is that the light from the object hitting the plate (or film) should be almost as bright as the reference beam. Making use of this set-up it is quite hard to adjust the relative intensities of these two sources of light, the reference beam tending to be much brighter than the objects' reflected light. To increase the amount of reflected light, choose objects which are either red or white in colour and place them as close to the photographic plate as you can without obstructing the reference beam. To gauge the relative intensity of the two sources of light place the piece of white cardboard (described above) in the plate holder and block off with a piece of black cardboard first the reference beam and then the objects' reflected light. You may then judge their relative intensities by eye. They should be roughly equal, although better results may be obtained if the light from the object is slightly dimmer than the reference beam. As in determining the exposure, best results are obtained by experimenting with different relative intensities but, once again, these are not too critical.

You may use only objects which will remain absolutely still during the exposure. Organic objects—e.g., living plants or even food—will tend to move during the exposure and be recorded either dimly or not at all. A simple test to see whether an object is moving is to observe the speckle pattern which is formed across its surface; if the pattern is changing or moving (in which case it appears as if ants were crawling across the surface), the object itself is moving and may not be holographed.

Having set up the apparatus so that the plate will be satisfactorily illuminated, you can start making holograms. Plates or film of the 10E75 type should be used. Place the laser's shutter in position, turn out the room lights, open the box of plates, remove one, place it in position with the emulsion side of the plate facing the objects and the reference beam, and carefully repack the remaining plates in the box. You may now make an exposure (by lifting the shutter) after which the plate must be put in a light-tight container and processed, as described above.

After processing the plate, remove the objects and place the plate in the holder as for exposing, with the developed emulsion side again facing the reference beam. By looking through the plate you should be able to see a 3D image of the object which apparently occupies the same position behind the plate as it did when the hologram was exposed. Great care must be taken when doing this that you don't look directly down the laser beam (reference beam) which is being reflected from the mirror. To view the real image, arrange the hologram as shown in fig. 18, and look in the direction indicated. Alternatively, you can put the plate in the holder the opposite way round (with the emulsion side facing towards you) and look through the hologram in the same direction as used to view the virtual image. The real image should appear to stand out in front of the plate and be pseudoscopic, although you may have difficulty focussing on it (it may appear slightly larger than the original object).

If your hologram is not successful it may be due to any of the following reasons: movement of part of the apparatus (in which case make all components more solid) or of the vibration-free table; incorrect exposure of the plate; incorrect development and processing of the plate (or film); or incorrect balance of light from object and reference beam.

If you have used shiny objects as the subject matter of your hologram (i.e., objects which reflect light specularly), you may obtain some spurious images upon reconstruction since the light reflected from them may be focussed to produce alternative reference beams to the one reflected from the mirror. Each of these beams forms a separate hologram and all of these are superimposed on the plate; when you illuminate the hologram with just the reference beam from the mirror all the images from the multiple hologram are reconstructed and they will not coincide with the intended image. The unwanted images may be larger or smaller than the original objects. When making holograms of shiny objects, better results may be obtained by placing a 'diffuser', such as a frosted glass screen, in the path of the illuminating beam, close to the objects.

Simple white light reflection hologram

To obtain best results when making reflection holograms you should use plates (or film) of the 8E75 type. The detail in the interference pattern of a reflection hologram is even finer than that of a transmission hologram, and 8E75 plates are designed to be able to record this. Because of this fine detail, reflection holograms are more susceptible to failure caused by movement of the apparatus than are transmission holograms. You may find that components used satisfactorily to make a transmission hologram of the type we have described are not solid enough for making a reflection hologram.

155

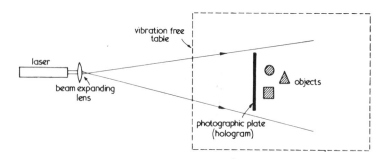

32 Arrangement for producing simple white-light reflection hologram (page 155).

A simple arrangement for producing reflection holograms is shown in fig. 32, and is attributed to the Russian scientist Denisyuk who in turn based his ideas on those of the Frenchman Gabriel Lippmann, who proposed a novel technique for colour photography in the early 1900s. It utilises the fact that the emulsion with which 8E75 glass plates are coated is quite transparent to light and will let it pass through. The expanded beam of the laser hits the plate and travels through to the other side; this constitutes the reference beam or wavefront. At the same time some of the light is reflected back towards the plate from the surface of the object; this constitutes the object wavefront and interferes with the reference wavefront to form an interference pattern throughout the emulsion, which becomes, on development, a white light reflection hologram.

Set up the components without the plate, as shown in fig. 32, making sure that the laser beam is wide enough to illuminate the whole of the plate when it is placed in position. With this arrangement the reference beam is almost certain to be much brighter than the object light, so choose white, silver or red objects and place them close to the plate. When you are satisfied that everything is ready place the shutter over the laser, switch off the lights, put the plate in position with the emulsion side facing the objects, make the exposure and remove the plate to process it as in Section I. The optimum exposure-time is about five seconds for a one milliwatt laser and ten seconds for a half milliwatt laser.

After processing the plate you may view the recorded image by reflecting white light produced by a small bright source off the surface of the hologram that was facing the laser; sunlight is ideal, but you could also use a spotlight, penlight (with a lens in the bulb) or the light from a slide projector. Look into the hologram and tilt it in different directions until you see the image apparently behind the hologram. This will appear green due to the emulsion shrinkage (see Chapter 4). If you cannot see anything ask a friend to have a look before giving up, since the images may be very faint.

If you fail to obtain an image then try making another hologram after reinforcing the various components, particularly the plate holder, and try using brighter objects (e.g., silver ones) placed closer to the plate—they may even be touching it. Also try using a different exposure-time, longer or shorter depending on the darkness of the first hologram.

The main disadvantage of this technique is that it is impossible to increase the relative brightness of the light reflected from the objects. After attempting the beam-splitter hologram in the next section you might like to try making a set-up for reflection holograms which uses a beam splitter, thus allowing the brightness of the objects to be increased with respect to the reference beam.

If you have 8E75 film from which the anti-halation layer has been removed you can fashion it into a cone to produce a cone-shaped hologram by first cutting it (in the dark) using a semicircular template and then taping it together. For stability, attach a heavy metal ring around the base, using transparent adhesive tape, and place the cone over the object to be hologrammed—for instance, a single die. If this is now illuminated from above by laser light a reflection hologram will be obtained just as with a flat glass plate. To direct the laser beam so that it comes from above, you must reflect it from a mirror supported vertically above the cone. The laser may be tilted upwards to accomplish this, provided it is stably supported in this position. After processing the film a bright white light placed directly over the cone will provide suitable illumination for the reconstruction of the image.

Transmission hologram with a beam splitter

We can effect a number of improvements on the set-up described on pages 152–5 by incorporating a beam splitter into the apparatus. By using a beam splitter we obtain an object-illuminating beam and a reference beam which both have a circular cross section. Also, we can arrange to have a wider angle between the reference beam and the light coming from the objects so that: (a) the objects may be placed closer to the photographic plate without obscuring the reference beam; and (b) there is less danger of accidentally looking down the reference beam when viewing the hologram in the reconstruction stage.

Along with these advantages there are a couple of disadvantages. The fineness of detail in the hologram depends on the angle between the objects and the reference beam, so that as the angle gets bigger the interference fringes in the hologram are closer together and the whole set-up is more prone to failure by vibration or movement in the apparatus. At very large angles the pattern may be so fine that the photographic plate is incapable of recording it satisfactorily. If the angle is much more than 45° the reconstructed image will appear quite dim. Also, depending on the set-up used, the light which is eventually reflected from the objects will have travelled a substantially longer path than the reference beam. If this difference in path lengths is greater than the coherence length of the laser (which is roughly equal to the length of the laser), then the set-up will not produce satisfactory holograms and only a small depth of the scene will be successfully recorded.

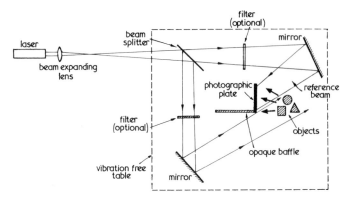

33 Arrangement for producing transmission hologram with beam splitter.

In the arrangement to be described here this difference is quite small and can, to a certain extent, be adjusted by repositioning the components.

The basic set-up is shown in fig. 33. The opaque baffle (which may be made from a piece of black cardboard) is included to prevent stray light hitting the plate from the beam illuminating the objects. Considerable care is needed to ensure the correct alignment of all the components. Firstly ensure that the laser beam is wide enough by the time it reaches the photographic plate to illuminate it uniformly across its surface (by placing a white plate-sized card in the plate holder). In order to obtain the most efficient reconstruction from the hologram you should contrive to make the angle between the objects and the reference beam as small as possible.

Alignment of the various components may be achieved quite simply by removing the beam-spreading lens and sending the narrow beam of the laser through the system. It should strike the beam splitter, each of the mirrors, and the photographic plate at the centre of their front surfaces. You can see if this is the case since the beam produces a spot of light at each surface that it hits (in fact you may see two beams reflected, side by side, from the beam splitter due to reflection both from the front and back surfaces, but these travel in the same direction and the alignment process is basically as described). The beam which travels to the objects from the beam splitter should hit the approximate centre of the object scene; but ensure that after replacing the beam spreader the objects do not obstruct the reference beam, so casting a shadow on the plate. Once again check that the plate will be uniformly illuminated across its surface.

You may now proceed to make holograms in the manner we have previously described. Again, experiment with a range of different exposure-times.

For reconstruction, best results will be obtained if you simply place the hologram in the single expanded beam from the laser, as shown in fig. 34. It should be at roughly the same angle to this beam as it was to the original reference beam. The real image may be seen by viewing the hologram as shown in fig. 18 or by turning the hologram around so that the side which had emulsion on it faces away from the laser beam (see page 154).

If your system fails to produce holograms, and you have tried all the remedies suggested earlier, then you may have to try to reduce the path difference between the two light paths. The path difference may be measured with a tape measure and, if necessary, changed by adjusting the positions of the components; for example, moving the photographic plate and the objects to the left in fig. 33 increases the distance the reference beam travels, while decreasing the total distance travelled by light reflected from the objects. This adjustment

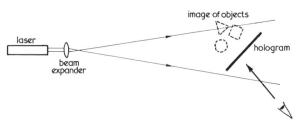

34 Simple viewing arrangement for transmission hologram of fig. 33.

necessarily changes the geometry of the system and you should once again check its alignment.

Another advantage of a system with a beam splitter is that we may, if necessary, balance the intensity of the two beams by placing a neutral grey (or even coloured) filter in one of the positions shown in fig. 33. This is useful if you are recording holograms of objects which are not very bright, since it enables you to ensure that the light reflected from the objects and the reference beam is of roughly the same intensity on reaching the photographic plate. This may be tested as before by putting a white card in place of the plate and alternately blocking off the two beams while watching the brightness of the light falling on the card.

This set-up is probably the best of those described in this chapter in terms of the quality and brightness of the holographic images produced. Unfortunately, you need a laser or another very monochromatic source to view the holograms. Try illuminating the hologram with the light from a slide projector with a colour filter fitted in the slide compartment— you should be able to see the image. Another commonly available source for demonstrating your holograms is a sodium street lamp, which produces a fairly pure yellow light. You should find that when the hologram is illuminated by different coloured sources that the image appears larger or smaller than the original object. The laser, however, remains the best source of light for illuminating your holograms to obtain optimum quality.

You may want to photograph your holograms. This will require exposure times of several seconds (or even minutes), so you must use a tripod to hold the camera. Remember to focus on the point behind the plate at which the object appears to stand. You will find that at high f-numbers the effect of speckle in the images is very noticeable, producing grainy photographs. By reducing the f-number you can get rid of this graininess but only by sacrificing depth of focus in the picture.

Section III   Further Experiments

Once you have made holograms using the arrangements described above you may want to try some experiments with them. Double-exposure 'trick' holograms (Chapter 4) are simple enough to make, but you must ensure that none of the components, particularly the photographic plate, are disturbed as you rearrange objects in the scene. The time for each of the exposures should be half that for a single-exposure hologram (one third if you make three exposures, etc.).

Double-exposure interferometry holograms (Chapter 6) can also be made with the set-ups described. Choose an object which can be held firmly and devise some means for giving it a very slight distortion. For instance, if you have a vice which may be mounted temporarily on the vibration-free table the object may be held in this and deformed slightly after the first exposure, by tightening the vice. A small plastic bottle is a very good subject for such a hologram.

As mentioned earlier, the beam which emerges from the laser and beam spreader often has an interference pattern already in it which becomes annoyingly superimposed on the hologram. The only adverse effect that this has is that you have to look through this pattern, which lies in the plane of the holographic plate, in order to see the image. You can get rid of this pattern and clean up the beam using a device known as a spatial filter, which consists of a tiny pinhole placed exactly at the focus of the beam-spreader lens. This filters out the unwanted light waves in the laser beam that are the cause of the interference pattern. These devices may be purchased ready-made, but they are very expensive, costing as much or more than the laser, and it is possible to make one yourself. The pinhole may be produced in a piece of aluminium foil by placing it on a piece of glass and pricking it with a needle or pin. The foil may then be mounted in a cardboard ring and the hole placed at the focus of the lens. This can be a very tricky and time-consuming operation, needing much patience, but the results should justify the effort. To make it easier, the pinhole and ring may be mounted in an adjustable (i.e., up-and-down, side-to-side) lens holder intended for use on an optical bench. Suitable 'optical posts' can be purchased for the laser so that it too may be mounted on the bench. After delicate adjustment of the pinhole you should find a position at which a beautifully 'clean' expanded beam is produced by the beam spreader and filter (place a screen in a suitable position to see this). Plate 23 compares filtered and unfiltered beams from a laser. Having adjusted the pinhole, be careful not to jolt it out of place, particularly when making an exposure using a cardboard shutter.

Another improvement you may like to make is in your developing technique. Use of fine-grain developer can improve the quality of your holograms, though experiments need to be made to determine the best developing time. You could also try bleaching your holograms. There are various techniques for doing this; for instance, Agfa-Gevaert recommend a process described in a leaflet available from them. The resulting phase holograms are brighter than unbleached holograms but often not of such good quality.

Having exhausted the suggestions given here for making holograms you may wish to refer to other texts on practical holography. These are listed in the Bibliography (page 166). Indeed, if you are contemplating going to the expense of buying a laser you might like to read one of these books first in order to get other ideas and a different view on the subject.

Ultimately you must rely on your own imagination to provide ideas for new holograms and different set-ups. Future developments in laser technology and the mass-production of lasers may reduce the cost of some of the more sophisticated lasers, enabling amateurs to purchase them and so to greatly extend the type of holograms which they can produce. Pulsed lasers would facilitate holographic portraiture, and three-colour lasers, such as the Argon-Ion laser, would enable

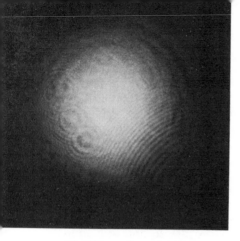

23 Effect of spatial filter on an expanded laser beam. These photographs show the spot produced when the laser beam hits a white wall (a) with no filter, and (b) with a spatial filter. Filtering removes the unwanted pattern in the beam which is otherwise superimposed on the hologram itself.

natural-colour holograms to be made. Strict safety regulations apply to these lasers at the present time, since they are higher-powered than the small helium-neon lasers we have recommended. In many parts of the USA it is necessary to pass special examinations before being licensed to use such lasers, and similar restrictions apply in other countries.

Amateur photographers who do their own developing and printing are familiar with the way in which time seems to fly past while one is engrossed in one's work. The same thing happens when making holograms. If, having read this chapter, you decide to try your hand then I wish you good luck and ask you always to remember the importance of safety when working with lasers—*never* look directly into the laser beam.

Useful Addresses

Agfa Gevaert Inc.
275 North Street
Teterboro NJ 07608
(201) 288-4100
holographic plates and film

Ealing Corporation
2225 Massachusetts Avenue
Cambridge MA 02140
optical equipment

Edmund Scientific Co.
101 E. Gloucester Pike
Barrington NJ 08007
(609) 547-8900
optics, lasers and holograms

Gallery 1134
1134 West Washington Blvd
Chicago IL 60607
(312) 226-1007
exhibitions, courses

Holos Gallery
1792 Haight Street
San Francisco CA 94117
(415) 668-4656
holograms, exhibitions

Interference Hologram Gallery
008-1179A King Street West
Toronto, Ontario M6K 3C5
(416) 535-2323
exhibitions, courses

Integraf
Box 586
Lake Forest IL 60045
(312) 234-3756
holographic plates and film

Metrologic Instruments Inc.
143 Harding Avenue
Bellmawr NJ 08030
lasers, hologram-making kits

Museum of Holography
11 Mercer Street
New York NY 10013
(212) 925-0526
holograms, books, exhibitions
publishes newsletter 'holosphere'

NRC Newport Corp
18235 Mount Baldy Circle
Fountain Valley CA 92708
(714) 963-9811
basic and advanced hologram kits
lasers, optics, plates and film

Spectra Physics Inc.
1250 West Middlefield Road
Mountain View CA 94042
lasers

Bibliography

Listed here are some of the texts and articles which I have found useful in inquiring about holography. Works of general interest are listed first, with texts relevant to specific chapters following, according to the surname of their authors. Not all of these sources are readily available; some can only be obtained from university libraries. Ask at your local library or bookseller; they may be able to help you. Notes are included describing some of these books and articles.

General

Caulfield, H.J., 'The Wonder of Holography,' *National Geographic*, March 1984. Up-to-date article; magazine cover features an embossed plastic hologram.

Caulfield, H.J., ed, *Handbook of Optical Holography*, Academic Press, New York, 1979 (hardback). An up-date on the material in Collier's book; technical.

Collier, R. J., and Lin, L. H., *Optical Holography*, Academic Press, New York, 1971 (hardback). A large book, considered by many to be the holographer's bible. Contains a comprehensive review of most of the holographic research undertaken before 1971. Technical account.

Jeong, T. H., *Study Guide on Holography*, 1975 (paperback). Non-technical holography textbook, theory, experiments to perform. (Available by post from T. H. Jeong, Box 586, Lake Forest, Illinois 60045.)

Okoshi, T., *Three-Dimensional Imaging Techniques*, Academic Press, New York & London, 1976 (hardback). A fascinating (technical) textbook which deals not only with holography but also with other recent 3D techniques descended from the stereoscopic principle (Chapter 1) which, because of their versatility, have wide applicability. Okoshi feels certain that these other techniques will be implemented in such applications as 3D television well before equivalent holographic methods.

Saxby, Graham, *Holograms*, Focal Press, London 1980 (hardback). Introductory level text, practical experiments, applications, mathematical theory.

Smith, H. M., *Principles of Holography*, Wiley Interscience, New York, 1969 (hardback). A technical textbook. The general historical introduction and Chapter One should appeal to the general reader.

Unterseher, F., Hansen, J., and Schlesinger, B., *Holography Handbook: Making Holograms the Easy Way*, Ross Books, Berkeley, California, 1982 (paperback). The most up-to-date practical handbook available, with simple and advanced techniques, description of artists' work, plans for home-built equipment.

About Light

Much of this material is covered in secondary-school science textbooks.

Gregory, R. L., *Eye and Brain*, World University Library, Weidenfeld and Nicolson, London, 1966 (hardback and paperback). A standard first-year undergraduate psychology text which is nevertheless a readable and fascinating account of the process of seeing.

See also: Gernsheim (1969).

The Laser

Brown, R., *Lasers*, Aldus, London, 1968 (hardback). A not too technical account of lasers and their applications.

See also: Kock (1969).

Applications of Lasers

Barkan, R., 'The Laser Goes Into Battle', *New Scientist*, 13 July 1972, p. 84.

See also: Holoco (1976), Brown (1968).

Principles of Holography

Gabor, D., 'A New Microscopic Principle', in the journal *Nature*, Vol. 161, p. 771 (1948). Gabor's original paper in which he described the process of holography.

Leith, E. N., and Upatnieks, J., 'Photography by Laser', *Scientific American*, June 1965. One of the first, and still one of the best, introductory articles on holography, addressed to the non-specialist. Unfortunately a little dated. Written by the pioneers of 3D laser holography.

Pennington, J., 'Advances in Holography', *Scientific American*, February 1968.

Phillips, N. J., and Porter, D., 'An Advance in the Processing of Holograms', *Journal of Physics (E)* (GB), volume 9, no. 8, pp. 631–4, August 1976.

Display Holography

Benthall, J., *Science and Technology in Art Today*, Thames and Hudson, London, 1972 (paperback). A fascinating account of all forms of technological art with a chapter devoted to holography, containing some interesting philosophical ideas about the subject.

Benton, S. A., 'Holography: The Second Decade', *Optics News*, Summer 1977. The author, inventor of the rainbow hologram, reviews recent developments in the field.

Benyon, M. 'Holography as an Art Medium', *Leonardo*, Vol. 6, pp. 1–9 (1973).

Hammond, A. L., 'Holography: Beginnings of a New Art Form or At Least of an Advertising Bonanza', *Science*, Vol. 180, p. 484 (1973).

Leith, E. N., 'White Light Holograms', *Scientific American*, October 1976.

Rogers, M., 'A State of the Art Report: Holography', *Rolling Stone Magazine*, 30 August 1973, p. 36.

Wuerker, R. F., and others, 'Holography in the Conservation of Statuary', *Studies in Conservation*, Vol. 18, pp. 49–63 (1973).

See also: Holoco (1976).

Holography as a Tool

Briers, J. D., 'Trends in Holography', *Physics Bulletin*, Vol. 27, p. 202 (May 1976). A review of some recent techniques.

Caulfield, H. J., and Lu, S., *The Applications of Holography*, Wiley Interscience, New York, 1970 (hardback). A short technical account of the applications, intended for the non-specialist.

Dudley, D. D., *Holography*: *A Survey*, Technology Utilisation Office, NASA, Washington, D.C., 1973 (paperback); available from Museum of Holography, New York (see addresses). Describes technical applications of optical, acoustic and microwave holography.

Greguss, P.: Professor Greguss is the inventor of 'bioholography', a theoretical model for information processes which occur in the brains of animals. His ideas are somewhat controversial and have not gained general acceptance among workers in the field. He has written several papers on the subject, such as 'Bioholography—a New Model of Information Processing', *Nature*, Vol. 219, p. 482 (1968); and, among the more controversial, 'What Makes the Beautiful Appear as Such? Scopes of Bionical Aesthetics' (in Hungarian) and 'Computer Simulation of Acupuncture Anaesthesia Based on a Bioholographic Model'. Nevertheless he is an acknowledged expert in the field of acoustic holography; he has edited a book, *Holography in Medicine*, IPC Science and Technology Press, Guildford, UK, 1973, available from the Museum of Holography, New York (see addresses) or from the publishers. The last eleven chapters are devoted to holographic brain/retina/hearing models, concepts in natural occurrences, etc.

Leith, E. N., and Upatnieks, J., 'Progress in Holography', *Physics Today*, March, 1972, p. 28. A review of holography and technical developments.

Lesem, L. B., Hirsch, P. H., and Jordan Jr., J. A., 'The Kinoform: A New Wavefront Reconstruction Device', *IBM Journal of Research and Development*, March 1969.

Metherell, A. F., 'Acoustical Holography', *Scientific American*, October 1969.

See also: Smith (1969), Collier and others (1971).

The Future of Holography

Bentov, Itzhak, *Stalking the Wild Pendulum*, pub. by E. P. Dutton, New York, 1977 (hardback).

See also: Caulfield (1970), Collier and others (1971), Benthall (1971).

Making your own Holograms

Dowbenko, G., *Homegrown Holography*, Amphoto, New York, 1978 (paperback). Contains practical details of many simple systems, particularly 'sandbox' technique.

Heumann, S. M., 'How to Make Holograms and Experiment With Them', *Scientific American*, February 1967, p. 122.

Hsue, S. T., and others, '360-Degree Reflection Holography', *American Journal of Physics*, Vol. 44, no. 10, p. 927 (1976).

Lehmann, M., *Holography—Technique and Practice*, Focal Press, London, and Hastings House, New York, 1970 (hardback). Explains advanced technique. Very useful if you have access to metalworking facilities.

Outwater, C., and van Hamersweld, E., *The New Guide to Practical Holography,* **ODCD Press, 22030 Grant Avenue, Torrance, CA 90503, no date, (paperback). A very good account of practical holography, worth a look if you are seriously considering making your own holograms. Videotape also available.**

Stirn, B. A., 'Recording 360-degree Holograms in an Undergraduate Laboratory', *American Journal of Physics*, Vol. 43, no. 4 (April 1975), p. 297. Describes simple method of producing cylinder transmission holograms which could be accomplished by amateur.

See also: Jeong (1975).

171

Acknowledgements

Of the many people who have provided help, counselling and advice during the preparation of this book, special mention should be made of: Julian Dale; Thelma Young and Dr Atkins of Bristol University; the staff at the Museum of Holography, New York; Jonathan Benthall; Harriet Casdin-Silver; Mr Croucher of Agfa-Gevaert; Nick Phillips; the staff and students of the Optics Section, Imperial College, my editor, Paul Barnett; my friends Over the Rainbow and in London; my typewriting mother and proof-reading family; the Star take-away, Earl's Court Road; and of course Dennis Gabor, for a novel technique of producing 3D images.

Several people very kindly gave me permission to reproduce photographs, and acknowledgements are as follows: **2**, courtesy of Bell Laboratories; **5**, courtesy Ralph Wuerker, TRW Systems Inc.; **7**, courtesy Newport Research Corporation; **8**, courtesy Nick Phillips, University of Loughborough; **9**, courtesy R. Rhinehart, Macdonnel-Douglas Corporation; **10**, courtesy Ralph Wuerker, TRW Systems Inc.; **11**, photograph courtesy Mike Sage; **12**, hologram by Multiplex Co., photograph courtesy MGM Films Inc.; **13**, copyright 1969 by International Business Machines Corporation and reprinted with permission; **14**, courtesy Stephen Benton, Herb Mingace and Will Walters; **15**, courtesy T.J. Kujawa, Apollo Lasers; **16**, hologram by M. Benyon; **17a**, hologram by H. Casdin-Silver in collaboration with S. Benton, photograph by Nisham Bichajian; **17b**, hologram by H. Casdin-Silver in collaboration with S. Benton, photograph by Scott E. Nemtzow; **18**, pendant courtesy R. Rallinson, Electric Umbrella Inc.; **19**, courtesy Newport Research Corporation; **20**, courtesy R. Wuerker, TRW Systems Inc.; **21**, courtesy R. Wuerker, TRW Systems Inc.; **22**, guitar by Martin Co., hologram by N.E. Molin and K.A. Stetson, 1969.

Earls Court,
June, 1978

Index

A separate index is given for the material covered in chapter 8.

173

INDEX

Index to Chapter 8; making your own holograms

INDEX